T0340943

The Secrets of Alchemy

SYNTHESIS

A series in the history of chemistry, broadly construed, edited by Angela N. H. Creager, Ann Johnson, John E. Lesch, Lawrence M. Principe, Alan Rocke, E.C. Spary, and Audra J. Wolfe, in partnership with the Chemical Heritage Foundation

The Secrets of
ALCHEMY

LAWRENCE M. PRINCIPE

The University of Chicago Press
Chicago and London

The University of Chicago Press, Chicago 60637
The University of Chicago Press, Ltd., London
© 2013 by The University of Chicago
All rights reserved. Published 2013.
Produced in the United States of America

22 21 20 19 18   10 11 12 13 14

ISBN-13: 978-0-226-68295-2   (cloth)
ISBN-13: 978-0-226-10379-2   (paperback)
ISBN-13: 978-0-226-92378-9   (e-book)
DOI: 10.7208/chicago/9780226923789.001.0001

Library of Congress Cataloging-in-Publication Data

Principe, Lawrence, author.
    The secrets of alchemy / Lawrence M. Principe.
        pages cm — Synthesis
    Includes bibliographical references and index.
    ISBN-13: 978-0-226-68295-2 (hardcover : alkaline paper)
    ISBN-10: 0-226-68295-1 (hardcover : alkaline paper)
    ISBN-13: 978-0-226-92378-9 (e-book)
    ISBN-10: 0-226-92378-9 (e-book)   1. Alchemy—History. 2. Alchemists.   I. Title.
II. Series: Synthesis (University of Chicago Press).
    QD13.P75 2013
    540.1'12—dc23

                                                    2012010213

♾ This paper meets the minimum requirements of the ANSI/NISO Z39.48-1992
(Permanence of Paper).

❈ CONTENTS ❈

WHAT IS ALCHEMY?

Although alchemy's glory days came to an end roughly three centuries ago, the Noble Art endures in many ways. The very word *alchemy* conjures up vivid images of the hidden, the mysterious, and the arcane, of dark laboratories and wizard-like figures bent over glowing fires and bubbling cauldrons. Today, most people have heard something about the Philosophers' Stone, the substance capable of turning lead into gold that was so eagerly sought by legions of alchemists. Indeed, an entire generation became acquainted with the stone and one of its supposed possessors, the medieval Parisian notary Nicolas Flamel, by means of the first of J. K. Rowling's wildly successful books: *Harry Potter and the Philosopher's Stone*. (Regrettably, American publishers corrupted the substance's ancient name into the meaningless "Sorcerer's Stone." Alchemy has not always gotten the respect it deserves.) The sixteenth-century Swiss alchemist Theophrastus von Hohenheim, better known as Paracelsus, recently found new life as "Hohenheim of Light" in the Japanese manga and anime series *Fullmetal Alchemist*, which makes copious if highly sensationalized use of alchemical concepts. Trading on the link between alchemy and transformation, many modern books have *alchemy* in their title, thereby renewing alchemy's modern presence virtually every year. Such titles range from Paul Coelho's 1988 best-selling novel *The Alchemist* to more trite borrowings of the term as *The Alchemy of Love* and *The Alchemy of Finance*, to the more imaginative *American Alchemy: The History of Solid Waste Management in the United States*. The alchemical theme of transformation is also responsible for the frequent appearance of the term in various self-help programs.

Besides these manifestations of variously transformed versions of alchemy, a perhaps surprising number of people throughout the world are continuing to search for practical metallic transmutation, despite rather discouraging prognostications from modern chemistry, often more or less in the same ways it was pursued centuries ago. Some such modern seekers—I know from personal experience—even hold positions in university departments. Alchemy thus continues to exist in a variety of guises and disguises.

But much of the modern world's familiarity with alchemy is more apparent than real. While the mystique of the subject naturally attracts interest, its inherent difficulty and complexity easily deflect attempts to understand it. Arriving at solid, satisfactory conclusions about alchemy can seem as difficult as finding the Philosophers' Stone itself. Alchemy's primary sources present a forbidding tangle of intentional secrecy, bizarre language, obscure ideas, and strange imagery. The alchemists did not make it easy for others to understand what they were doing. Secondary sources about alchemy, whether books or websites, are frequently even more problematic, for they soon plunge the reader into a maze of conflicting claims and contradictory assertions. The historically informed works readily available today range from excellent scholarly publications (which naturally presuppose considerable expertise) to introductory but now outdated overviews.[1] Far outnumbering works by historians, however, are those by an assortment of popular writers, occultists, enthusiasts, and a few hucksters that recapitulate a variety of clichés, misconceptions, historical errors, and baseless opinions, rather than presenting the current state of knowledge about the subject. Most such books link alchemy in various ways—both favorable and unfavorable—to religion, psychology, magic, theosophy, yoga, the New Age movement, and, perhaps most often, to loosely defined notions of the "occult." Without guidance, it thus proves extremely difficult for even the most intrepid seeker to emerge from such a labyrinth with any clear or sound conclusions about the true nature of alchemy.

So what *is* alchemy? Who were the alchemists, and what did they believe and do? What were their goals, and what did they accomplish? How did they envision their world and their work, and how were they seen by contemporaries? These are the main questions I explore in the following pages.

My goal is to provide a reliable guide to the various secrets of alchemy. A comprehensive history of the subject not only would run to unreadable length but also would be premature, since scholars still have much to learn about it. I present instead only an introduction that can serve as a solid foundation for further inquiry. My chief motivation, then, in writing this book has been to make some of the enormous wealth of recent discoveries about alchemy accessible to a wider audience. While alchemy has always been considered secret and privileged knowledge, perhaps the best-kept alchemical secret of our day is how radically our understanding of the subject has changed during the last forty years. Alchemy is now a hot topic among historians of science. Books and manuscripts that have lain unread for centuries are now being read again, and their contents more accurately understood in historical context. We are learning more about alchemy every day. Yet much of this new information has remained inaccessible to most readers, because it is published in specialist literature and in multiple languages—more often than not in languages other than English. The result has been that most popular writing about alchemy repeats the same mistaken notions over and over again, perpetuating errors that were satisfactorily and convincingly corrected in the scholarly literature as much as eighty years ago. I believe that interested readers deserve much better.

I wrote *The Secrets of Alchemy* to function on two levels. In the main body of the text, I keep the nonspecialist, the general reader, and the student in mind. No prior knowledge of alchemy or specialized knowledge of the history of science is necessary to understand it. Some familiarity with chemistry will help in chapter 6, but is not crucial. For readers who want to delve more deeply into one or more aspects of the subject, however, I have supplied extensive endnotes directing them to more advanced treatments. These notes are intended to function as a categorized (but not exhaustive) guide to the most reliable current scholarship on the subject as well as to sound editions of primary texts. I have not been so pedantic as to list *every* source available on each topic; I have instead chosen only the best and most pertinent. My apologies to scholars whose relevant works I have not yet encountered; I would be delighted to receive references or offprints.

I have strenuously avoided making this book a *Who's Who* of alchemy. Many practitioners of the discipline, including some important ones, receive only passing mention or even none at all—a fact that might disappoint some readers. I chose instead to focus on a small number of important characters, each of whom represents a major trend or feature within alchemy. Readers will thus gain a deeper familiarity with the thinking of a few foundational figures who can act as reckoning points within the long alchemical tradition, rather than coming away with a superficial overview of many characters.

Alchemy's Periodization and the Structure of This Book

Historians of science customarily divide the history of Western alchemy into three main chronological periods: the Greco-Egyptian, the Arabic, and the Latin European. The Greco-Egyptian (and later, Byzantine) period, which stretches from the third to the ninth century, set alchemy's foundations and established many features that would characterize it for the remainder of its life. The Arabic or Islamic period (eighth to fifteenth century) sought out this Greek heritage and then massively augmented it with fundamental theoretical frameworks and a wealth of practical knowledge and techniques. Thus when alchemy arrived in medieval Europe, it came as an Arabic science, its lineage signaled by the Arabic definite article *al-* affixed to the word itself. It was thereafter in Europe that alchemy saw its greatest flowering and largest following. After its establishment in the High Middle Ages (twelfth to fifteenth century), alchemy achieved its golden age during the early modern period (sixteenth to early eighteenth century), an era widely known as the Scientific Revolution. Not only was the alchemy of this period the most developed and diverse of all, but we possess vastly more sources dating from this time than from earlier ones.

To these three eras of the standard periodization should be added a fourth, which stretches from the eighteenth century to the present day. It is to this (ongoing) period that we owe influential "revivals" and radical reinterpretations of earlier alchemical traditions, several of which generated lively cultural and intellectual movements of their own. This period should be treated as a significant part of the full history of alchemy. It is also to this period that we owe most of the misconceptions about pre-eighteenth-century alchemy that remain widespread. Consequently, it is better to examine the origins of these depictions of alchemy, and to situate them in their due historical contexts, so that they do not distract from our efforts to gain a more historically accurate depiction of alchemy as it existed before the eighteenth century. To that end, revealing the surprising (and surprisingly late) origins of many ideas about alchemy widely held today is sufficiently important to warrant violating chronological order. Therefore, chapters 1 through 3 cover Greco-Egyptian, Arabic, and medieval Latin alchemy, respectively, but chapter 4 jumps over alchemy's golden age of the sixteenth and seventeenth centuries to treat the eighteenth-century "end" of alchemy and the subsequent era of reinterpretations and revivals. Chapter 5 resumes the chronological sequence by exploring alchemy in the early modern period.

Topics akin to those pursued in Western alchemy were also subjects of early inquiry further East—that is to say, in India and China. However, the Indian and Chinese material is not covered here. The major reason is simply that we do not yet have a sufficiently comprehensive or accurate understanding of it. Furthermore, when previous treatments of alchemy attempted to combine Eastern and Western alchemy into a single narrative, the result was usually more confusion, not greater clarity. For example, an ahistorical conflation of Chinese and Western alchemy spawned the popular, but erroneous, notion that European alchemists sought an "elixir of immortality." Although Western practitioners did seek medicines that would *extend* life, the search for earthly immortality through alchemy was a uniquely Chinese goal. Eastern and Western pursuits and practices do bear certain resemblances, but they are embedded in such widely divergent cultural and philosophical contexts that trying to squeeze them into a single narrative damages the uniqueness of each one. The Western term *alchemy* might even prove to be a misleading label for the Eastern practices known more correctly as *waidan* and *neidan*. In any event, meaningful historical linkages between Eastern "alchemy" and Western alchemy remain unidentified (although contact within the Islamic world is certainly plausible), and so it is unwise to

assume or assert such linkages in the current absence of clear and compelling historical evidence. The two "alchemies" of East and West are, at
least at this point in time, better treated as separate entities.[2]

The last three chapters of this book recount aspects of alchemy's great
flowering in sixteenth- and seventeenth-century Europe. Chapter 5 gives
an overview of early modern alchemical theory and practice, and outlines
its terminology and goals in making metals, medicines, and other arcana.
Chapter 6 confronts the difficult question of what alchemists were actually doing in their laboratories. I approach this question by two complementary routes: the textual and the experimental. The first and more
traditional route involves deciphering the bizarre language and imagery
that early modern alchemists routinely used to conceal their knowledge
and activities. The second and more novel route involves replicating the
deciphered alchemical processes in a modern laboratory to see and do
what early modern alchemists saw and did, and to test the correctness of
the textual interpretations. Chapter 6 both explains step by step how to
understand enigmatic texts and images purporting to teach the preparation of the Philosophers' Stone, and reveals the actual chemical basis for
the secretly encoded processes. The results are often very surprising.

Alchemy's place in early modern Europe extended far beyond the
confines of smoky laboratories; it diffused itself through a wide swath
of contemporaneous culture. Artists, poets, humanists, playwrights, devotional writers, theologians, and many others borrowed from and com-
mented on alchemy. Their works offer additional perspectives on the
Noble Art. Additionally, some ways of thinking natural to alchemists
illustrate profound differences between the ways early modern people
saw and thought about their world and the ways we (or at least most of
us) do today. The study of alchemy therefore opens a window onto a remarkable and meaning-rich vision of the world that has largely been lost
today. This vision was by no means unique to alchemists; it was common
throughout European culture of the day. Failing to understand that vision means failing to understand not only alchemy but our predecessors
as a whole, and indeed diminishing ourselves by allowing a crucial part
of Western heritage to fall into oblivion. The seventh and final chapter
presents these wider worlds of alchemy.

The study of alchemy—and of the past in general—brings us into contact with the diverse ways thinkers of other times and cultures conceived
of the world, how they answered questions the world posed to them, and

how they made use of the powers and riches of that world. This why we study history: to see, at least for a time, with the eyes of others, and to be enlightened and enriched by the fresh (but ancient) ways they might see even the most common and neglected of things. In this regard, alchemy still has much to teach.

ORIGINS

Greco-Egyptian *Chemeia*

To locate the origins of alchemy, we must travel back to Egypt in the first centuries of the Christian Era. This place was no longer the Egypt of the far more ancient pharaohs and pyramid builders but a cosmopolitan, Hellenized civilization. Egypt had come under the influence of Greek culture following its conquest by Alexander the Great during his vast military campaigns of 334–323 BC. Even after Egypt's absorption into the Roman Empire in the first century BC, its dominant culture and language remained Greek. By the first century AD, its major city, Alexandria (founded in 331 BC and named for Alexander himself), had become a vibrant crossroads for cultures, peoples, and ideas. From this Eastern Mediterranean melting pot, the earliest surviving chemical texts, and even the origin of the word *chemistry* itself, date.

Many technical operations fundamental for alchemy had been developed well before its emergence. The smelting of metals such as silver, tin, copper, and lead from their ores had been practiced already for four thousand years. The making of alloys (such as bronze and brass, both

alloys of copper) and various techniques for metallurgy and metalworking had been developed to a fairly high degree. In Egypt, artisans had
devised an array of processes for making and working glass, producing
artificial gems, compounding cosmetics, and creating many other commercial products in what might be called an ancient chemical industry.[1]
Generations of workshop laborers had devised and refined these techniques, with the tricks of the trade passed down from father to son, from
master to apprentice.

The Technical Literature: The Papyri and Pseudo-Democritus

The earliest documents that scholars routinely attach to the history of
alchemy bear witness to this technological and commercial background.
These precious and unique texts, written in Greek on papyrus, date from
the third century AD. They were discovered in Egypt in the early nineteenth century and now reside in museums in Leiden and Stockholm;
hence they are called the Leiden and Stockholm Papyri.[2] They contain
about 250 practical workshop recipes. These recipes fall into four chief
categories: processes relating to gold, to silver, to precious stones, and
to textile dyes, all costly articles of luxury and commerce. Significantly,
most of the recipes deal with how to make *imitations* of these valuable
substances: coloring silver to look like gold, or copper to look like silver;
making artificial pearls and emeralds; and coloring cloth purple using
cheaper imitations of the extravagantly expensive imperial purple dye
made from murex snails. Since the Papyri also contain a series of tests to
determine the purity of various metals, both precious and common, it is
evident that the original users of these formulas clearly understood the
difference between genuine and imitation articles.

We can get a better sense of what these craftsmen were doing by trying to follow in their footsteps. The eighty-seventh recipe in the Leiden
Papyrus describes the "discovery of the water of sulfur." The ancient
text's directions are these: "Lime, one dram; sulfur, previously ground,
an equal quantity. Put them together into a vessel. Add sharp vinegar
or the urine of a youth; heat from underneath until the liquid looks like
blood. Filter it from the sediments, and use it pure."[3] The ingredients of
this recipe are simple, clearly identifiable, and readily obtainable, so we
can replicate the process today. After the ingredients are mixed (I found
that urine works better than vinegar, by the way) and boiled gently for

about an hour, an orange-red and unpleasantly scented liquid results. Although the Leiden Papyrus does not say *how* to use the liquid, we can guess. When a polished piece of silver is dipped into it, the metal quickly becomes tawny, then golden, then coppery, then bronzy, purple, and finally brown. Impressively, the shiny brilliance of the metal remains undiminished by the color changes until the very end, and the color and sheen remain stable for long periods of time. With a little practice and careful control of the temperature and the length of time the metal is left in the solution, I succeeded in making silver look astonishingly like gold (see plate 1).[4]

The color changes result from the formation of extremely thin layers of sulfides on the metal surface, owing to the action of calcium polysulfides present in this "water of sulfur." To be sure, similar compositions are still used occasionally today for patinating metal objects (in other words, producing changes to their surface color).

Recipes such as this one provide a necessary background to the emergence of alchemy, but they are not themselves, strictly speaking, alchemical. Alchemy, like other scientific pursuits, is more than a collection of recipes. There must also exist some body of theory that provides an intellectual framework, that undergirds and explains practical work, and that guides pathways for the discovery of new knowledge. Alchemy moreover was to be about more than making look-alikes of precious substances.

It is important to realize that these papyri are the *only* original documents currently known to survive from the Greco-Egyptian period. Despite the many books about alchemy that we know were written during that time, the only surviving testimony of that distant era comes in the form of corrupt anthologies—that is, collections of excerpts copied from original texts that are now lost. These anthologies—collectively called the *Corpus alchemicum graecum*—were compiled by Byzantine scribes, and the earliest of them dates from a time long after Greco-Roman Egypt had itself become a faded memory. The oldest surviving copy dates from around the start of the eleventh century, and many of its pages are missing. It contains excerpts from about two dozen books dating from the second to the eighth century, and is now preserved in Venice. This manuscript, called Marcianus graecus 299, is supplemented by a few later manuscripts now in Paris and elsewhere that contain additional texts or alternate readings. While priceless to scholars, these collections represent only a frustratingly slim remainder of alchemy's foundational epoch.[5] Equally problematic is the fact that the Byzantine compilers chose to copy what *they* thought was important—which could be neither representative of the original texts nor what the original authors themselves would have considered crucial. Hence, the overall picture of what Greco-Egyptian alchemists thought and did is skewed by the way their writings were excerpted centuries later.

The earliest text within the *Corpus alchemicum graecum* dates from about the late first or second century AD. It carries the title *Physika kai mystika*, and the text we possess is fragmentary. Its author is named as Democritus; but he is certainly not, as is sometimes claimed, the ancient philosopher of the fifth century BC famous for his notion of atoms.[6] The title, which may have been given to it much later, is often translated as *Physical and Mystical Things*. Although that might *look* like a reasonable rendering of the Greek, it is misleading. A better translation is *Natural and Secret Things*. The Greek word *mystika* did not refer in ancient times to what we today call mystical, that is, something having a special religious or spiritual meaning, or expressing a personal experience of the ineffable. Instead, it simply meant things to be kept secret.[7] Calling this text *Physical and Mystical Things* immediately suggests that the author was describing both material and spiritual things, but this is not the case. The *Physika kai mystika* records workshop recipes similar to those of the Leiden and Stockholm Papyri. In fact, it uses the same fourfold division of processes into those for gold, silver, gems, and dyes. This similarity of format suggests that a whole tradition of practical recipe books once existed in which this division was standard. For pseudo-Democritus, these processes are *mystika*, that is, *secret*, because they are lucrative artisanal processes—trade secrets, if you will.

Nevertheless, the text also contains an account of how the frustrated author, unable to carry out his craft adequately because his master had died before teaching him the necessary techniques, tried to contact the deceased. The attempt was only half successful. The master's shade spoke only to say that he was not allowed to relay information freely across the gulf that now divided him from the living, and that "the books are in the temple." A little later, a pillar in the temple suddenly opened up to reveal a hidden niche containing a terse expression of the master's secret knowledge: "Nature delights in nature, nature triumphs over nature,

nature masters nature."[8] (This is not the only tale of alchemical secrets suddenly revealed in a place of worship.) This repetitive and rather obscure phrase is used like a refrain throughout the recipes of the *Physika kai mystika*. Whatever meaning we attach to this tale of discovery, the recipes themselves remain straightforward and practical, with no trace of the mystical (in a modern sense) or the supernatural.

The Birth of Alchemy

The recipe literature such as the Papyri and the *Physika kai mystika* aims to imitate or extend precious materials. But probably during the third century AD, a crucial juncture in the emergence of alchemy was reached. At some point—no texts survive to inform us of exactly how or when this first happened—the idea of actually making *real* gold and silver emerged. This development would have seemed reasonable enough from the point of view of a worker at that time. If the water of sulfur can tinge the surface of silver to look like gold, why shouldn't there be some way to tinge it through and through—even more than that, to give silver not only the color of gold but *all* the properties of gold? The process for making gold is called *chrysopoeia*, from the Greek words *chryson poiein* (to make gold), and it is accompanied by the less common (and less lucrative) *argyropoeia*, the making of silver. The general process of transforming one metal into another is called *transmutation*.

From this point onward, alchemists had a coherent goal toward which to strive with both head and hand. They would pursue a great many things besides chrysopoeia, but the making of gold and silver remained one of the central goals of what would come to be called the Noble Art. The authors of the earliest alchemical treatises borrowed techniques, processes, and tools from a wide variety of contemporaneous artisans, yet they saw themselves as a group distinct from those artisans.[9] Thus, both alchemy and alchemists acquired an independent identity in the third century.

The birth of alchemy required the union of two traditions: the practical artisanal knowledge exemplified in the recipe literature, and theoretical speculations about the nature of matter and change present in Greek natural philosophy: What is matter? How does one thing change into another? A Greek speculative tradition centering on these questions stretched back for some seven hundred years before the emergence of alchemy. Such questions preoccupied the earliest Greek philosophers, known collectively as the pre-Socratics. The first thinker generally cited in this tradition is Thales of Miletus (sixth century BC), who claimed that all the different substances around us are really modifications of a single primordial substance that he identified as water. Many other thinkers followed Thales with their own ideas. Democritus and Leucippus (fifth century BC) proposed the concept of invisibly small *atomoi* (atoms), from which everything is composed. Empedocles (circa 495–435 BC) attributed the origin of natural substances and their transformations to four "roots" of things he called fire, air, earth, and water. These four combine in various ways and separate under the influence of forces he called love and strife. Perhaps most prominently of all, Aristotle (384–322 BC) devoted substantial attention to the nature of matter and change, devising theories and ways of thinking that would prove highly influential and fertile for further investigations.

All these Greek philosophers endeavored to explain matter's hidden nature and to account for its unending transformations into new forms. Most of them embraced the idea that beneath the constantly changing appearances of things, there existed some sort of a stable, unchanging substrate. The notion that a single ultimate substance lies beneath all material things is known as *monism*. For Thales, this ultimate substance was water; for Democritus, imperishable atoms; for Aristotle, what he called "first matter" or "prime matter" (*prōton hylē*). Empedocles' four elements, strictly speaking, represent a position of *pluralism*, since he implied that more than one kind of ultimate matter exists, but he nevertheless maintained the idea of a constancy beneath change. So far as we know, however, these natural philosophers had only a secondhand acquaintance with the practical knowledge of the crafts.

In the cosmopolitan crossroads of Greco-Roman Egypt, the two streams of craft traditions and philosophical traditions coexisted. Their merger—probably in the third century AD—gave rise to the independent discipline of alchemy. The intimate mingling of the two traditions is evident in the earliest substantial texts we have about chrysopoeia. These writings come from a Greco-Egyptian alchemist who would be revered as an authority for the rest of alchemy's history, and the first about whom we have any reasonably substantial or reliable historical details: Zosimos of Panopolis.

Zosimos of Panopolis

Zosimos was active around 300 AD.[10] He was born in the Upper Egyptian city of Panopolis, now called Akhmim. We know that he was not the first chrysopoeian, because his writings refer to earlier authorities, and even to rival "schools" of alchemical thought that had already developed by his time. (Of these other schools we know absolutely nothing save what he writes in criticism of them.) Zosimos is thought to have written twenty-eight books about alchemy; alas, most of what he wrote is now lost. We have only scraps: the prologue to a book titled *On Apparatus and Furnaces* (sometimes called the *Letter Omega*, under which letter it was once classified),[11] several chapters from other works, and scattered excerpts. Some of Zosimos's writings are addressed to Theosebeia, a woman who seems to have been his pupil in alchemical matters, although whether she was a real person or a literary device we will never know for sure. Despite the fragmentary nature of what survives and the difficulty in interpreting it, these writings provide the best window we have onto Greek alchemy. Surprisingly, these early texts establish many concepts and styles that would remain fundamental for much of later alchemy.

Zosimos's orientation toward a central goal (metallic transmutation), his insightful engagement with the practical problems in reaching it, his search for the means of surmounting these problems, and his formulation and application of theoretical principles clearly underscore his writings as something new and significant. While earlier texts are recipe miscellanies, Zosimos's texts witness a coherent program of research that draws on both material and intellectual resources. He describes a wide array of useful apparatus—for distillation, sublimation, filtration, fixation, and so forth—in great detail.[12] Many of these instruments are adapted from cooking utensils or items used in perfumery or other crafts. Zosimos did not devise all these instruments himself, indicating how developed practical chrysopoeia must already have become by the start of the fourth century AD. The writings of his predecessors form a key resource for him, and he cites them frequently. One of the most prominent authorities is named Maria—sometimes called Maria Judaea or Mary the Jew—and Zosimos credits her with the development of a broad range of apparatus and techniques. Maria's techniques include a method of gentle, even heating using a bath of hot water rather than an open flame. This simple but useful invention preserved the legacy of Maria the ancient alchemist, not only for the rest of alchemy's history, but even down to the present day. It is her name that remains attached to the *bain-marie* or *bagno maria* of French and Italian cookery.

Several of the pieces of apparatus Zosimos describes—for example, one called the *kerotakis*—are designed to expose one material to the vapors of another. Indeed, he seems particularly interested in the action of vapors on solids. This interest is partly grounded on practical observations. Ancient craftsmen knew that the vapors released by heated *cadmia* (or calamine, a zinc-containing earth) could turn copper golden by transforming it into brass (an alloy of zinc and copper). The vapors of mercury and arsenic whiten copper to a silvery color. Perhaps knowledge of these color changes induced Zosimos to seek analogous processes that would bring about true transmutations. Guiding theories are certainly discernible in his writings. This is a crucial point to stress. Today there is a common misconception that alchemists worked more or less blindly—stumbling about mixing a little of this and a little of that in a random search for gold. This notion is far from the truth; already with Zosimos we can identify *theoretical principles* that guided his practical work, as well as *practical observations* that supported or modified his theories. Many theoretical frameworks for alchemy would develop in various times and places, and these frameworks both supported the possibility of transmutation and suggested avenues for pursuing it practically.

In the case of Zosimos, not enough of his work survives to map out his thinking fully. Yet it is clear that he viewed the metals as composed of two parts: a nonvolatile part that he calls the "body" (*sōma*) and a volatile part that he calls the "spirit" (*pneuma*). The spirit seems to carry the color and the other particular properties of the metal. The body seems to be the same substance in all metals; in one fragment Zosimos appears to equate it with the liquid metal mercury. Thus, the identity of the metal is dependent on its spirit, not its body. Accordingly, Zosimos uses fire—in distillation, sublimation, volatilization, and so on—to separate the spirits from the bodies. Joining separated spirits to other bodies would then bring about transmutation into a new metal.

Across the gulf of ages, Zosimos's observant, active, questioning mind makes itself apparent. In one passage, he notices the disparate effects of sulfur vapor on different substances, and expresses his astonishment that while the vapor is white and whitens most substances, when it is absorbed by mercury, which is itself white, the resulting composition is yellow. Always ready to criticize his contemporaries, Zosimos chides them by saying that "they should inquire into this mystery first of all."[13] He likewise expresses his surprise that when the vapor of sulfur turns mercury into a solid, not only does the mercury lose its volatility and become fixed (that is, nonvolatile), but the sulfur also becomes fixed and remains combined with the mercury.[14] Zosimos's observation is now recognized as a basic principle of chemistry: when substances react with one another, their properties are not "averaged," as they would be in a mere mixture, but instead completely changed. Clearly, Zosimos was a careful observer who thought deeply about what he witnessed experimentally.

Zosimos calls transmutation the "tingeing" of metals, and uses the word *baphē*, from the verb *baphein*, which means "to dip" or "to dye"; he likewise calls a transmuting agent a "tincture," that is, something able to tint or color. These word choices signal the connection of his ideas to the recipe literature, which was primarily concerned with coloring metals, stones, and cloth to produce precious (or apparently precious) objects. Accordingly, the "water of sulfur" reappears prominently in Zosimos, but now with strikingly new meanings. It is no longer a simple composition for bringing about superficial changes but rather some putative substance able to bring about real transmutation—and consequently something eagerly sought and eagerly hidden.

Here an almost ubiquitious feature of alchemy appears: *secrecy and the hiding of names*. Zosimos delights in playing with the name of this substance. Thanks to an ambiguity in the Greek language, in some contexts the name can mean either "water of sulfur" or "divine water." In some places he intends the name to mean a transmuting agent, while in others he is clearly talking about the simple lime-sulfur composition of the recipe literature.[15] In yet another place he describes it as "the silvery water, the hermaphrodite, that which flees without ceasing . . . it is neither a metal, nor a water always in movement, nor a solid body, for one cannot grasp it."[16] In this case his riddle for "divine water" seems to describe mercury, presented as the basic substrate for all metals. Elsewhere, the same term seems to have yet other meanings. In point of fact, in a Zosimos text just recently identified, the Egyptian admits freely that alchemical writers "call a single thing by many names while they call many things by a single name."[17] He notes that the production of transmuting "waters" is "the manifest secret, that which is studiously hidden."[18] The moderate level of secrecy encountered in the earlier recipe literature thus becomes more intense and more self-conscious with Zosimos. Such secrecy would wax and wane in intensity but never disappear for the rest of alchemy's history.

To promote such secrecy, Zosimos employs a technique that would become typical for alchemical authors: the use of *Decknamen*, a German term meaning "cover names." These *Decknamen* function as a kind of code. Instead of using the common name for a substance, the alchemical writer substitutes another word—usually one that has some link, literal or metaphorical, with the substance intended. There is already some hint of this technique in pseudo-Democritus, where he uses the adjective *our* to specify a substance other than that usually meant by a common term; for example, he uses "our lead" to mean the mineral antimony (stibnite), a substance that shares some properties with lead. *Decknamen* serve a dual purpose: they maintain secrecy, but they also allow for discreet communication among those having the knowledge or intelligence to decipher the system. They simultaneously conceal *and* reveal. Consequently, *Decknamen* have to be *logical*, not arbitrary, so that they can be deciphered. If *Decknamen* could not be deciphered by readers, then total secrecy would be the result; and if the intent were to conceal information entirely, it would be far simpler for alchemists to have written nothing at all.

The encoding of information does not stop with simple replacements of the names of substances, not even in Zosimos. Perhaps the most famous fragments of the Panopolite are sometimes (and misleadingly) called his "Visions." Three fragments describe a series of five "dreams" separated by periods of waking. These dreams involve an altar shaped like a chemical vessel, various men of copper, of silver, and of lead, their violent dismemberment and death, and Zosimos's conversations with them. Much ink has been spilled trying to explain what these texts really mean. Regardless of the varied answers that have been offered over the past century or so, Zosimos himself tells us that they are allegorical descriptions of practical transmutational processes. In other words, the actors, places, and actions described are personified *Decknamen* woven into a coherent and extended narrative. Such allegorical language would remain a common feature of alchemical writing, and become especially prominent in works by European practitioners starting in the fourteenth century.

Zosimos calls his dream sequence a "prologue" intended to help the reader unveil the "flowers of speech" (*anthē logōn*) that follow. In the text as we have it today, only one practical process follows, but it appears that originally there were many more, now lost.[19] In another place, Zosimos writes clearly that after "awaking" from a dream, he "understood very well; those who busy themselves with these things [the events in the dreams] are the liquids of the metallic art."[20] In the book *On Sulphurs*, Zosimos uses a simile that compares the transmutation of lead into silver to a tormented man who becomes king; this image, which the text clearly links to a practical process, is very similar to those expressed in Zosimos's second "dream."[21]

Some modern writers have read various mystical or psychological meanings into Zosimos's allegorical accounts, but in so doing they have largely ignored their context—both within the corpus of his writings and within his cultural milieu. Zosimos clearly states that his "dreams" have a technical meaning in the context of the transmutation of metals—the primary topic of his texts. Some scholars have even proposed plausible interpretations of the "dreams" in terms of the Panopolite's alchemical theories and laboratory practices.[22] It is certainly possible that Zosimos did in fact dream (or daydream) about the work in which he was so deeply engaged; many readers have probably had similar experiences of work-related matters reexpressing themselves in strange dreams. But it is more probable that Zosimos composed these "dreams" explicitly, much like a fiction writer works, thus creating a self-consciously allegorical "prologue" for one of his practical treatises. This practice harmonizes well with his routine use of secrecy, and in fact, immediately after reciting one of these "dreams," he declares axiomatically that "silence teaches excellence," as if to explain his own relative silence and to advise an analogous silence for his readers.[23] The use of dreams as a literary device was an established and popular practice in Zosimos's day, and placing information into the form of a dream gives it a certain cachet—an air of authority and a tone of revelation.

Yet showing that the core meaning of Zosimos's "dreams" lies in practical alchemical operations does not mean we can ignore their broader cultural context. Zosimos surely drew upon his own experience and knowledge of contemporaneous religious rites for imagery to use in this allegorical sequence. His language of altars, dismemberment, and sacrifice surely reflects something of late Greco-Egyptian temple practices.

This recognition brings up a huge point for the entire history of science: how do practitioners' philosophical, theological, religious, and other commitments manifest themselves in the study of the natural world, whether in alchemy or elsewhere? Such studies—be they alchemical or modern scientific—do not occur in a cultural vacuum, nor are practitioners somehow insulated from the conceptions, interests, and ways of thinking of their particular time and place. Chapter 7 deals with the inseparability of such matters from alchemy and indeed from all scientific pursuits more generally. For now, it suffices to take one last illustrative look at Zosimos.

There is undoubtedly a link between Zosimos and Gnosticism. Gnosticism was a diverse grouping of religious movements of the second and third centuries AD that stressed the need for revealed knowledge (*gnōsis*) to achieve salvation.[24] This salvific knowledge included the realization that man's inner being was of divine origin but had become imprisoned in a material body. Knowledge was necessary to overcome man's ignorance (or forgetfulness) of his origins, enabling him to begin liberating himself (that is, his soul) from subjection to the body and its passions, and to the material world and the evil forces that govern it. The Gnosticism widespread in Zosimos's Greco-Egyptian milieu surfaces clearly in two places in his writings. One is the prologue to his *On Apparatus and Furnaces*, and the other is the fragment called the "Final Account."[25] The question is how and to what extent Gnostic ideas play a role in Zosimos's alchemical ideas.

In the first text, Zosimos rails against a group of rival alchemists who criticize *On Apparatus and Furnaces* as unnecessary. He counters that they think this way only because they are using phony tinctures (transmuting agents) whose apparent success is actually the result of spiritual beings called daimons.[26] The daimons trick these errant alchemists into believing that their preparations work, and as a result they claim that the specific equipment, materials, and procedures stipulated by Zosimos are not needed for success. The daimons thus use these false tinctures to manipulate their ignorant possessors, thereby keeping them under daimonic sway and subjected to Fate (an evil force to be rejected). What true alchemists seek, Zosimos declares, are tinctures that are purely "natural and self-acting," bringing about transmutation by the operation of their natural properties alone.[27] To prepare these true, natural tinctures, the right apparatus and the right ingredients and processes are absolutely necessary.

To drive home his point about the baleful results of allowing oneself to fall under the sway of daimons, Zosimos then gives a Gnostic account of the Fall of Man—how the original human being was deceived by maleficent spirits into being embodied as Adam. Zosimos reveals a Christian form of Gnosticism by recounting how Jesus Christ provided human beings with the knowledge needed for salvation, namely, the need to reject their "Adam" (the material body) in order to ascend again to their proper divine realm. Human imprisonment and its attendant evils thus arose in the first place from daimonic deception, just like that which now causes the errant alchemists to reject Zosimos's book. Surely, these bad alchemists are making their own circumstances worse by blindly continuing to be duped rather than liberating themselves from daimonic control. Zosimos's critical prologue must have originally provided an appropriate introduction to his (now lost) text about the furnaces and apparatus necessary for preparing a true transmuting tincture.

Does Gnosticism express itself visibly in Zosimos's alchemical theories or practices? Possibly. Given the Gnostics' fondness for casting their tenets into myth format, we could wonder if Zosimos's choosing to put alchemical processes into an allegorical dream sequence arises from the same tendency to mythologize doctrines—Gnostic or alchemical. Additionally, Zosimos's guiding theory of the twofold nature of metals (body and spirit) and the practical need to free the active, volatile soul from the heavy, inert body in order to achieve transmutations seems to parallel Gnostic views—and some other contemporaneous theological views—of man's divine soul as being trapped in a material body, and the consequent need to free it. For a Gnostic (or a Platonist, for that matter, and Zosimos wrote about Plato as well), human individuality and personality are found in the soul, not the body. In the same way, the metals draw their particular nature and identity from their *pneuma*, not their *sōma*.

We completely miss the fullness and multivalent complexity of premodern thought if we dissect it into modern categories. Zosimos had no reason to isolate his philosophical or theological commitments into special categories separated from the balance of his thought. Today there is a tendency to imagine that such "mixing" (it is mixing only from our perspective) somehow impedes rational and clearheaded work on practical matters, yet this is not only a modern prejudice but also far from true. Zosimos's methods—like anyone else's—of thinking about, conceiving, and interpreting his work could not help but be influenced by, and draw on, the totality of the way in which he conceived of the world as a whole. Thus, it is incorrect to say that alchemy for Zosimos was itself a religion, and an exaggeration to say that his alchemy was Gnostic. Yet it is equally wrong to imagine that Zosimos could (or should) "turn off" his ways of thinking, his mental landscape built upon contemporaneous Gnostic, Platonic, and other commitments, when at work on practical alchemical processes. Even modern scientists cannot do that, although some of them convince themselves that they can (perhaps under the trickery of a daimon named Pure Objectivity).

Before we leave Zosimos's time and place, there is one more piece of context to add. If scholars are correct to date Zosimos's activity to around 300 AD, then he witnessed not only Emperor Diocletian's violent suppression of a rebellion in Egypt in 297–98 but also the attempted destruction of alchemy's literary heritage by the same emperor. It is reported that Diocletian ordered all "books written by the Egyptians on the *cheimeia* of silver and gold" to be burned. The source, an account of the martyrdom of Christians during Diocletian's persecutions, claims that this measure was taken to prevent the Egyptians from amassing enough wealth to rebel again.[28] However, if indeed this book burning took place as reported, it may have been related to Diocletian's empire-wide monetary reforms, which included the replacement in 295–96 of Egyptian provincial coins (minted at Alexandria) with standard Roman currency.

The third century AD witnessed a steady monetary collapse for the Roman Empire. Mints increasingly debased the currency by striking coins containing less and less precious metal, thus widening the gap between the coins' face value and their intrinsic worth. The amount of silver in the coin called the *antoninianus*, for example, dropped from 52 percent to less than 5 percent. Many issues of bronze coins were given a superficial silver (or merely silvery) coating to make them appear to be worth more than they really were. Diocletian's solution (which ultimately proved unsuccessful) was to issue new coinage.[29] Since the Egyptian books often described means of mimicking precious metals, hiding the debasement of alloys, or—in the ideal case—producing new gold and silver, it seems that these sorts of processes would be the last thing a ruler intent on monetary stabilization would want to have around, especially in the hands of a rebellious province of the empire. Significantly, a substantial number of late antique coins made of imitation precious metal have recently been identified, and the composition of some of them is strikingly similar to what would be produced following the recipes in the Papyri and pseudo-Democritus.[30] If the fear of counterfeiting and currency debasement lay behind Diocletian's decree, it would be the first in a long line of concerns over the value of currency that resulted in proscriptions against alchemy. The imperial edict banning books about *cheimeia* might also provide some of the background for the enhanced level of secrecy apparent in Zosimos's writings.

Whether or not this last suggestion is correct, one feature of this account remains: it is one of the earliest usages we have of a term—*cheimeia*—from which the words *alchemy* and *chemistry* derive. It is now time to say something about these two words. As with so much of alchemy, many unreliable claims have been made about their origin. This situation dates to the alchemists themselves, who loved to indulge in drawing fanciful etymologies in order to make various claims about their discipline. A common practice in antiquity was to trace the name of a thing to that of a mythical founder—hence Rome draws its name from the mythical Romulus, for example. Zosimos refers to an early alchemist named Chēmēs or Chymēs, and in another passage claims that the art was initially revealed by an angel in a book titled *Chēmeu*.[31] Zosimos undoubtedly drew the germ of this notion from the apocryphal Hebrew *Book of Enoch* (or 1 Enoch), wherein fallen angels teach the productive arts to mankind. But even modern texts about the history of alchemy or chemistry often present unlikely origins. One popular notion is that *chemistry* derives from the Coptic word *kheme*, meaning "black," alluding to the "black land," Egypt, in reference to the color of Nile silt. There is some support for this notion, since the first-century-AD writer Plutarch notes that *chēmia* was an old name for "Egypt."[32] Hence, according to this theory, *chemistry* would literally mean "the Egyptian art." Less plausibly, others have linked this derivation to the "black stage," a crucial step toward effecting transmutation, or to the imagined nature of alchemy as a "black art."

But the word more likely has a Greek origin, given that Greek was the language both of the earliest alchemical texts and of literate Greco-Roman Egypt. The "chem" of *alchemy* and *chemistry* very probably derives from the Greek *cheō*, which means "to melt or fuse." *Cheō* also gives rise to the Greek word *chuma*, which signifies an ingot of metal. Since most of the early chemical practices involved the melting or fusing of metals, this etymology certainly seems the most plausible and reasonable. The Greek word for the subject is then *chemeia* or *chumeia*, literally an "art of melting [metals]." (A predominantly Greek etymology does not, however, rule out a double meaning that draws also on the Coptic root.) By the way, the use of the word *alchemy* in referring to the Greco-Egyptian period could be seen as an anachronism, since that word is an Arabized form of the older Greek term—the "al" of *alchemy* is simply the Arabic definite article. (So what Zosimos and his contemporaries practiced should perhaps be called "chemy" . . .) But more on terminology later.[33]

Later Alexandrian and Byzantine Authors

Several Greek texts about *chemeia* dating from after the time of Zosimos down to the eighth century survive.[34] Most are commentaries on earlier material, and as is the case with so much of early alchemy, several of their authors await further and more careful study. One important development within this material is a greater melding of the practical with the theoretical and philosophical. From Olympiodoros, a writer of the sixth century AD, we have a fragmentary commentary on a now-lost work of Zosimos. This Olympiodoros may very well be the philosopher of the same name who wrote commentaries on Aristotle. He followed the lead of earlier Greek thinkers—such as Thales—who sought to identify a universal material from which everything is made. Olympiodoros reorients this idea of a common material substrate to speak of a common "matter of metals," which, by being receptive to a variety of different qualities, gives rise to the various metals. Thus, transmutation would be accomplished by reducing a metal to its "common metallic matter" and then introducing the qualities of the desired metal. This idea of a common metallic matter subject to interchangeable sets of qualities seems a continuation of Zosimos's division of metals into "body" and "spirit." Interestingly, Olympiodoros also justifies the use of allegory in place of plain language in alchemy by noting how Plato himself used the same literary device when teaching his most important points.[35]

Stephanos of Alexandria, a Neoplatonic philosopher, commentator, astronomer, and scholar, wrote an alchemical work titled *On the Great and Sacred Art of Making Gold*, which has recently been dated to 617. In this book he explicitly applies ideas from Plato, Aristotle, and other notable Greek philosophers to alchemy.[36] Unlike Zosimos, however, neither Olympiodoros nor Stephanos seems to have been interested in practical work. Alchemy did not constitute their main interest; they were philosophical thinkers first. Accordingly, chrysopoeia was for them a philosophical issue, and perhaps we might think of them—at least from what we know presently—as armchair alchemists. Nevertheless, their application of Greek philosophical thought, especially regarding matter, to alchemy continued the construction of an increasingly sophisticated theoretical framework for chrysopoeia. Such developments were significant not just in themselves, but also because these later versions of alchemy would be inherited by the Arabic world.

An often-reproduced image that comes from Marcianus graecus 299 is probably an emblematic expression of the philosophical principle upon which so much Greek alchemical theory and practice is based. This figure is known as the *ouroboros*, a serpent swallowing its tail (fig. 1.1). Interpretations of this simple but arresting image vary widely. But the

Figure 1.1. The *ouroboros* from Marcianus graecus 299, fol. 188v. Reproduced in Marcellin Berthelot, *Collection des alchimistes grecs* (Paris, 1888), 1:132.

inscription within it—ONE THE ALL (*hen to pan*)—directs us again toward ancient Greek philosophical notions about a single material that serves as the underlying substrate for all substances. Clearly, this principle undergirds the idea of alchemical transmutation: one thing can be turned into another because at the deepest level they are really the same thing. Thus, as things appear to pass away and new things come to be, there is a sense in which they remain always the same: one thing is all things, all things are one thing. Thus, the serpent *ouroboros*, like the sum total of material substances, continuously consumes itself and produces itself from itself, remaining constant even while perpetually destroying and regenerating itself.

One other development is worth mentioning before departing the Greek-speaking world for the Arabic: new names for a specific substance that would bring about transmutation. In Zosimos, this substance is one of several things he meant by the phrase "water of sulfur." Another term he uses is *xērion*, which originally meant a medicine in the form of a powder to be sprinkled on wounds. This term may have been chosen for its relation to the word *pharmakon* (drug, salve, poison), occasionally used by pseudo-Democritus for various substances able to color metals. But the term *xērion* suggests another parallel, namely, that just as medicine heals and improves sick human beings, *chemeia* heals and improves base metals by the use of its own "medicine," the *xērion* or transmuting agent. This powerful agent of transmutation would acquire a new and more enduring name that appeared no earlier than the seventh century: *hō lithos tōn philosophōn*, the Philosophers' Stone. Discovering how to prepare that "stone which is no stone" would become the alchemists' paramount goal.[37]

TWO

DEVELOPMENT

Arabic *al-Kīmiyā'*

Alchemy developed extensively during its Arabic period, roughly 750 to 1400, augmented in every respect by new theories, concepts, practical techniques, and substances. Centuries of cultivation in the Islamic world produced a massive body of knowledge across the sciences, medicine, and mathematics that would earn the awe and admiration of medieval Europeans when they first encountered it in the twelfth century. Yet although medievals recognized the wealth and importance of Arabic scholarship, that esteem gradually eroded in later generations, until the contributions and even the names of influential Arabic authors became confused, forgotten, or even suppressed. Thus, despite the importance of this period for alchemy—and for the entire history of science—our knowledge of it remains very incomplete. Historians have had to rediscover the primary sources of Arabic alchemy. Only at the end of the nineteenth century did scholars begin to study Arabic alchemical texts again. Strikingly enough, we owe part of this renewal of interest to the chemist Marcellin Berthelot (1827–1907), the same person responsible for the publication of the Greek *Corpus alchemicum*.[1]

Since that time, many questions have been addressed, many gaps in our understanding filled, and many mysteries solved, but much more still awaits attention. Even for the most important Arabic authors, only a few texts have been edited, and fewer translated. Much-needed new scholarship has been stymied by the inherent complexity of the manuscripts and their loss through war and carelessness, as well as by regional political and economic situations that prevent free access to archives. Perhaps the most challenging problem, however, is the very small number of historians of science with linguistic skills in Arabic, and the yet smaller subset of these with an interest in alchemy.

The Transmission of Knowledge from Greeks to Arabs

In the mid-seventh century, shortly after the beginnings of Islam, Arab armies surged out of the Arabian Peninsula in all directions—north into Palestine and Syria, east into Persia, west across North Africa, and finally into Spain and even France. Most important for the story of alchemy is the Arab conquest of the Byzantine lands in the Eastern Mediterranean. In 640, the city of Alexandria was conquered and Egypt annexed to the Islamic Empire. There and in other formerly Byzantine holdings in the Middle East, the nascent Muslim world came into close contact with Greek ideas and culture. This intercultural contact strengthened in 661, when Mu'āwiyah, the second caliph (successors of the prophet Muhammad acting as leaders of Islam) of the Umayyad dynasty, established his capital at Damascus, in the heart of what just thirty years earlier had been Byzantine land. Thus, although the Umayyad caliphs were Muslim Arabs, their subjects were largely Byzantine Christians. The new Muslim rulers were skilled in warfare but not in running an empire, so they needed to employ experienced Byzantines as administrators, architects, and planners. This sociopolitical situation offered ample opportunity for the newly arrived Arabs to learn Greek ideas. Thus, a "translation movement" began, slowly and haltingly under the Umayyads, but greatly accelerated under their successors, the 'Abbasids, who moved the Islamic capital east from Damascus to the new city of Baghdad, which they founded in 762. There a host of translators labored to render hundreds of Greek books into Arabic: the writings of Aristotle and Plato, the mathematics of Euclid, and the medicine of Galen and

Hippocrates, as well as practical treatises dealing with technology, mechanics, and, of course, *chemeia*.[2]

We used to think we knew exactly how Greek *chemeia* first established itself in Arabic culture as *al-kīmiyā'*. The story begins engagingly enough with intrigue and murder at the Umayyad court in Damascus. Khālid ibn-Yazīd (died 704) was a young Umayyad prince, grandson of the caliph Mu'āwiyah. When Khālid's father died in 683 while besieging Mecca during a civil war, Khālid's elder brother succeeded to the caliphate, but died the next year at the age of twenty-two—and possibly not of natural causes. Because of Khālid's youth, the caliphate was then given to a relative by the name of Marwan, with the condition that Khālid would succeed him. But Marwan then married Khālid's widowed mother, promised the line of succession to his own sons, and declared Khālid a bastard. Khālid's mother's response was to smother her new husband with a pillow while he slept (some sources say she poisoned him). Given such a loving family, Khālid fled to Egypt. There, to put his lost caliphate behind him, the young prince began to study Greek learning, and found alchemy most to his liking. In some versions of the story, he encountered "Stephanos the elder," presumably the author Stephanos of Alexandria mentioned in chapter 1. Stephanos taught Khālid and translated alchemical books into Arabic for him. In other versions of the story, Khālid's instruction came instead from a Christian monk named Marianos. Accounts disagree on whether this monk was Greek or Roman, and whether or not he lived as a hermit in Jerusalem. In any event, Marianos had studied alchemy in Alexandria, supposedly under the tutelage of Stephanos, and shared that knowledge—including how to prepare the Philosophers' Stone—with Khālid. The prince himself then wrote several alchemical works to preserve the instructions he had received.

Khālid's books, and his status as "the first [Muslim] for whom medical, astronomical, and chemical writings were translated," are already recorded in a tenth-century Arabic source, as is the Christian monk Marianos.[3] Marianos's books are known today both in Latin translation and in Arabic.[4] Unfortunately, this tidy and engaging tale is pure fiction.[5] The books bearing the names of Marianos and Khālid ibn-Yazīd are actually compositions dating a century or more after the lifetimes of their reputed authors.

Yet there are some consolations for those who like the story. It remains plausible—although there is no clear evidence for it yet—that Egypt was a site for the transmission of the first alchemical texts to the Arabic world, even if Khālid was not involved (the transfer probably began to take place sometime after his death in 704). As for Marianos, Greek knowledge probably did come to Arabic readers at first through the intermediacy of Christian clerics; there are several well-attested examples of such transmission.[6] But the historical existence of Marianos is unlikely. Nevertheless, although this fictional seventh-century monk did not have the distinction of being the first to transmit Greek alchemy to Arabic readers, he would have the honor of being the first bearer of alchemy to another eager readership some five hundred years later. Under the Latinized name Morienus, he will reappear soon.

Without the convenient tale of Khālid and Marianos, the early assimilation of Greek alchemy into the Arabic world during the 700s remains obscure. What little we know of that early period is dominated by treatises written under the names of prominent Greeks. Zosimos's name was used, naturally enough, but so were the names of more famous individuals who never wrote a word about alchemy, such as Socrates, Plato, Aristotle, and Galen. At present we cannot tell if these texts are original Arabic compositions, translations of now-lost pseudonymous Greek works, or some combination of the two.[7]

Hermes and the Emerald Tablet

This early period of pseudoepigraphical Arabic works produced what would become perhaps the most revered and best-known text related to alchemy: the *Emerald Tablet*, attributed to the legendary figure Hermes. Hermes, called Trismegestus from the Greek words meaning "the Thrice-Greatest," is a complex layering of Greek and Egyptian mythological and heroic figures. The writings connected with his name are known collectively as the *Hermetica*, and comprise a diverse jumble of dozens of texts of Greco-Egyptian origin. Many are philosophico-theological of a Neoplatonic character, and date from the first to the fourth century AD. Others are astrological, technical, or magical, and some of these latter date to the first century BC. All these Hermetic texts were well known in late antiquity. None of them, however, bears any clear relation to alchemy.[8]

Yet Zosimos cites a "Hermes" as an authority. More significantly, by the tenth century in the Islamic world, Hermes had grown into the founder of alchemy, a native of Babylon, and the author of a dozen more-or-less alchemical works.[9] His fame and stature continued to grow thereafter. In the Latin West, his renown increased to the point that he was hailed as a contemporary or even a predecessor of Moses and a divinely inspired pagan prophet who foretold the advent of Christ. As a result, Hermes is the first and most prominent prophetic figure depicted in the late fifteenth-century pavement of the Cathedral of Siena in Italy. In Europe, Hermes likewise retained his position as alchemy's founder, to the extent that the term *Hermetic Art* became synonymous with alchemy/chemistry. As a result of the constantly developing myth of Hermes, the *Emerald Tablet*—though merely one paragraph long—developed into a foundational text for many alchemists, both Arabic and Latin. It was subjected to myriad lengthy analyses by dozens of authors, including Isaac Newton.[10]

The exact origin of the *Tablet* remains obscure. Most evidence indicates that it was written centuries after the bulk of the philosophical or technical *Hermetica*, and that it is an original Arabic composition dating from the eighth century. No Greek precursor or any earlier Greek citations of it have been located despite exhaustive searches.[11] It first appeared appended to a work which itself has complex and obscure origins, the *Book of the Secret of Creation* (*Kitāb sirr al-khalīqa*) by one "Balīnūs," an early ninth-century author writing in Arabic under the name of the much earlier Greek author Apollonios of Tyana.[12] Balīnūs's work is itself a pastiche; newer materials are combined with an earlier Syriac text by a priest named Sajiyus of Nablus, which itself incorporates yet older Greek material. How exactly the *Tablet* fits into this muddle remains unclear.[13] Nevertheless, it seems safe enough to doubt the veracity of the account told in the *Book of the Secret of Creation* that the text was discovered, written in Syriac on a tablet of green stone, clenched in the hands of an ancient corpse buried in a subterranean sepulcher hidden beneath a statue of Hermes Trismegestus.[14]

What is clear is that the *Emerald Tablet* never disappeared from view for long thereafter. It reappeared with various wordings and in various textual locations. Trying to make sense of it, however, occupied and frustrated a long line of would-be interpreters. The text is short enough that an early version can be presented here in its entirety.

Truth! Certainty! That in which there is no doubt!

That which is above is from that which is below, and that which is below is from that which is above, working the miracles of one thing.

As all things were from one. Its father is the Sun and its mother the Moon. The Earth carried it in her belly, and the wind nourished it in her belly, as Earth which shall become fire.

Feed the Earth from that which is subtle, with the greatest power.

It ascends from the Earth to the heaven and becomes ruler over that which is above and that which is below.[15]

We can see how readers convinced of the antiquity and importance of this text could have spent many sleepless nights striving to discern its meaning. The relationship between the celestial world (the macrocosm, "that which is above") and the terrestrial world (the microcosm, "that which is below") seems to be clear enough. There also seems to be a reference to monism ("all things were from one"), akin to the meaning of the *ouroboros*. But what is the "it" whose "father is the Sun"? Generations of alchemists believed that "it" was the Philosophers' Stone, the agent of metallic transmutation, and thus that the *Tablet* contained secret information about how to prepare that precious substance. But what are the Sun and Moon? Dry and wet principles perhaps? Gold and silver? Where is the Earth's belly? How and with what subtle thing are we supposed to feed the Earth? It is far from clear that the unspecified "it" has any relation to the stone or to practical alchemy at all. The mysteries of the *Emerald Tablet*—both its origins and its meaning—are not likely to be resolved anytime soon.

There is a curious anecdote dating from the tenth century about early Arabic interest in alchemy. The historian Ibn al-Faqīh al-Hamadhānī describes a visit made by 'Umāra ibn-Hamza, the ambassador from Caliph al-Mansūr, to the Byzantine emperor (probably Constantine V) sometime between 754 and 775.[16] According to this account, the emperor showed 'Umāra several impressive wonders of Constantinople, including storehouses filled with bags of white and red powders. In the sight of the Muslim ambassador, the emperor ordered a pound of lead to be melted, and threw a small amount of the white powder into the crucible. The lead was immediately turned into silver. Then a pound of copper was melted, and upon adding a pinch of the red powder it was transmuted into gold. 'Umāra reported this wondrous feat to al-Mansūr, who then suddenly developed an interest in alchemy and ordered Greek works about it translated into Arabic. Whether this is a faithful account of what 'Umāra reported or a later rewriting of events, at least the timing is right. For it was in fact under al-Mansūr, the clever founder of Baghdad (reigned 754–75), that the translation movement of scientific and medical works into Arabic began in earnest. This anecdote is of special importance, because it is an early account of *two* transmuting agents, a white one for making silver and a red one for making gold. These two forms of the Philosophers' Stone would become standard parts of transmutational alchemy.

Jābir and the Jābirian Corpus

The obscurity that clouds our understanding of the early transmission of alchemy to the Muslim world does not last long; it is replaced by confusion. For now we come to a person who played as large a role in Arabic alchemy as Zosimos did in the Greco-Egyptian—one Jābir ibn-Ḥayyān. Or, to speak more accurately, several Jābir ibn-Ḥayyāns. Or perhaps none at all. A persistent problem facing historians of alchemy is figuring out if an author really is who he says he is, and if he lived when and where he claims. Anonymity, pseudonymity, secrecy, mysteries, false trails, and subterfuge fill the entire subject from beginning to end. In the case of Jābir, disagreements about both author and his writings began shortly after his reputed lifetime and continue to the present day. As in the case of Khālid and Marianos, things in alchemy are often not what they seem.

Traditional biographies record that Abū Mūsā Jābir ibn-Ḥayyān was born about 720 at Kufa, an ancient town south of Baghdad. In his youth he learned alchemy first from Harbī the Himyarite (who died in 786 at the advanced age of 463) and then from a Christian monk often identified as a disciple of Marianos. (Getting suspicious yet?) Jābir's most important master, however, was a looming figure in Islamic religious history: the sixth Shi'ite Imam, Ja'far al-Ṣādiq (700–765). Jābir attributes his knowledge directly to Ja'far, whose closest and most beloved disciple he claims to have been. Some sources claim that Jābir himself became an imam and/or a Sufi. After Ja'far's death, Jābir went to Baghdad and became close to the rich and powerful Barmaki family, who introduced him to the court of Caliph Hārūn al-Rashīd (of *A Thousand and One Nights* fame, and who reigned 786–809), for whom Jābir wrote an alchemical volume. Jābir's date of death is variously given as 808, 812, or 815.

Doubts about this account were already circulating in the tenth century. The Baghdadi bookseller Ibn al-Nadīm reports that "many scholars and elders among the booksellers have affirmed that this man, Jābir, did not exist at all."[17] But al-Nadīm rejects this claim on the grounds that no one would write so many books—he lists about three thousand—and put another's name on them. (Authoring three thousand books is not as absurd as it sounds, since these "books" [*kutub*] were akin to chapters or short essays of a few pages, not whole volumes.) Other Arab writers voiced doubts; the fourteenth-century literary historian Jamāl al-Dīn Ibn Nubāta al-Miṣrī asserts that the consensus in his day was that Jābir was a pseudonym used by several different authors.

Arguments about Jābir raged anew in the early twentieth century as historians of science were rediscovering Arabic alchemy. But it was Paul Kraus, a scholar of immense erudition and linguistic prowess, who wrote the decisive work about Jābir.[18] Kraus concluded that the traditional biographies placed Jābir more than a century too early. As evidence, he noted that certain Greek sources to which Jābir refers were not available in Arabic in the eighth century, and that some of Jābir's basic ideas come from that crucial encyclopedic work the *Book of the Secret of Creation*, which was composed between 813 and 833, *after* the dates usually given for Jābir's death. Moreover, the bulk of Jābir's writings show the influence of a Shi'ite movement dating to the end of the 800s.

Kraus also argued that many authors were responsible for Jābir's three thousand books, and that these had been composed over the course of a century. The earliest, *The Book of Mercy* (*Kitāb al-raḥma*), was written in the mid-ninth century. This book, he postulated, excited interest among Shi'ite alchemists, who either wrote companion pieces for it or interpolated their own ideas into other preexisting texts to produce new "Jābir" writings around the end of the ninth century. This group also invented a connection for Jābir to their own historical master, Shi'ite Imam Ja'far al-Ṣādiq (who does not appear in the earlier *Book of Mercy*).[19] Further works were added to Jābir's name until the second half of the tenth century. Thus, the Jābirian corpus represents the evolving production of a "school" of alchemists.[20] There might have been an actual Jābir ibn-Ḥayyān somewhere in all of this, but not one with the biography or bibliography claimed for him. Therefore, when I write "Jābir," it is hereafter shorthand for "the authors of the writings under the name Jābir."

The Mercury-Sulfur Theory of the Metals

Jābir's writings contain practical information about processes, materials, and apparatus, along with a wealth of theoretical frameworks. The most enduring contribution connected with him is the *Mercury-Sulfur theory of the metals*. Presented in the *Book of Clarification* (*Kitāb al-īḍāḥ*), this theory has a long history before Jābir. It derives ultimately from Aristotle (384–322 BC), who postulated the existence of two "exhalations" that emanate from the center of the earth; one is dry and smoky, the other wet and steamy.[21] Underground, these exhalations condense and produce stones and minerals. Jābir's immediate source is not Aristotle, however, but rather that crucial early ninth-century work, the *Book of the Secret of Creation* by Balīnūs.[22] Zosimos's interest in sulfur vapor and his notion that mercury is the common "body" of metals may also play an intermediary role between Aristotle and Balīnūs.

The Mercury-Sulfur theory in Balīnūs, as recapitulated in Jābir, states simply that all metals are compounds of two principles called Mercury (akin to Aristotle's moist exhalation) and Sulfur (akin to the smoky exhalation). These two principles, condensed underground, combine in different proportions and degrees of purity to produce the various metals. As Jābir writes,

> The metals are all of the substance of quicksilver coagulated with the mineral sulphur that rises into it in a smoky exhalation of the earth. They differ only in their accidental qualities which depend upon the different forms of sulphur which enter into their composition. For their part, these sulphurs depend upon the different earths and their exposure to the heat of the sun. The most subtle, pure, and balanced sulphur is the sulphur of gold. This sulphur coagulates quicksilver with itself in a complete and balanced manner. On account of this balance, gold withstands fire, remaining unchanged in it.[23]

Thus, gold results from the perfect combination of the finest Sulfur and Mercury in exact proportions. But when the Mercury or Sulfur is impure, or the two are mixed in the wrong ratio, baser metals are produced. This theory provides the theoretical foundation for transmutation. If all metals share the same two ingredients and differ only in the relative proportions and qualities of those ingredients, then purifying the Mercury and Sulfur in lead and adjusting their ratio should produce gold.

Two points need to be stressed about the Mercury-Sulfur theory of the metals. First, until the eighteenth century, only seven metals were recognized. Two were considered noble (gold and silver), and five were considered base (copper, iron, tin, lead, and mercury).[24] The distinction between "noble" and "base" depended not only upon the relative monetary value of the metals but also upon their intrinsic beauty and their ability to resist corrosion. Second, the metallic principles Mercury and Sulfur were not necessarily identical with the common substances called by those names. These names were attached to the condensed exhalations by analogy with the properties of the common substances. Arabic alchemists knew very well that when they combined common mercury and sulfur in their workshops, they obtained cinnabar (mercuric sulfide), not a metal. The Jābirian corpus even gives a clear recipe for making cinnabar by dripping mercury into molten sulfur.[25]

The Mercury-Sulfur theory proved astonishingly long-lived. It was accepted (in various forms and to various degrees) by most chemical workers until the eighteenth century, almost a thousand years after it was first proposed. This longevity reflects both its conceptual utility and the fact that observable phenomena seem to support it. Some metals, such as iron and copper, burn vividly when finely powdered and dropped into a fire, and in doing so often emit a sulfurous smell. This simple observation supports the idea that they contain some kind of a flammable, sulfur-like substance. Tin and lead melt extremely easily, and when melted are visually indistinguishable from common mercury, thus suggesting that they contain a great deal of some sort of a liquid ingredient similar to mercury. A smaller proportion of this liquid ingredient could explain why iron and copper are so hard to liquefy—they are too "dry." Likewise, tin and lead are soft and pliable, while copper and iron are hard and brittle, as if the former had too much liquid in their composition, and the latter too little (think about, for example, potter's clay mixed with too much or too little water). Finally, the rusting or corrosion of base metals implied that they were "falling apart," decomposing because their ingredients were poorly or weakly combined, unlike the stronger, more stable composition of the noble metals gold and silver.

Jābir's Transmuting Elixirs: Aristotelian Qualities, Galenic Degrees, and Pythagorean Numbers

If transmutation requires just a simple adjustment of proportions, how would this process be carried out in practice? Jābir's practical guidelines start by drawing upon two concepts from Greek natural philosophy. The first is Aristotle's notion of the *four primary qualities* and their relation to the *four elements*. Aristotle stated that the most fundamental (hence "primary") qualities of any thing are hot, cold, wet, and dry. When pairs of these qualities are joined to matter, the four elements—fire, air, water, and earth—result.[26] The combination of hot and dry produces fire, cold-wet gives water, cold-dry produces earth, and hot-wet yields air (see fig. 2.1). Aristotle thought of these four elements as abstract principles of compound bodies, not actual substances that could be put into a jar and labeled. Jābir, however, was more a chemist than Aristotle; in the Jābirian corpus, these elements have concrete existence as isolable, physical substances.

When almost any organic substance—for example, wood, flesh, hair, leaves, eggs—is gradually heated, various materials are driven off sequentially by the heat, leaving behind a solid residue. Jābir interprets this practical experiment as the separation of a compound substance into its component elements. The "fire" distills off as a flammable and/or colored substance, the "air" as an oily one, and the "water" as a watery one; the "earth" remains behind as the residue. Once these elements are separated by distillation, Jābir wants to break them down further by removing one of their two qualities. According to Aristotle, water is the combination of the two qualities wet and cold with matter, so Jābir orders his readers to distill the separated water repeatedly from something with the quality of "dry"—he suggests sulfur. By repeated distillation, the dryness of the sulfur destroys the wetness of the water, so the alchemist is left with something *simpler* than an Aristotelian element: matter endowed with coldness alone. Naturally, the water's manifest properties change as its wetness is removed, and Jābir claims that after repeated treatment the

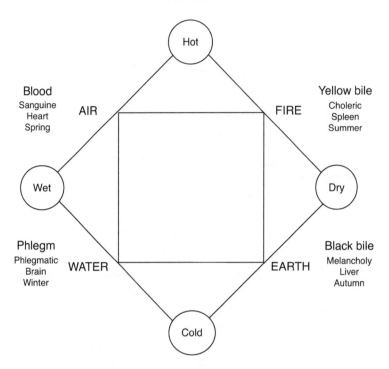

Figure 2.1. A schematic diagram showing the origin of the four elements from
the four primary Aristotelian qualities, and the relation of the four humors,
complexions, organs, and seasons to the four elements.

water turns into a glistening white solid similar to salt. Each element is
to be treated chemically to produce four substances, each one of them
bearing just one of the primary Aristotelian qualities.[27]

Once the four single-quality substances are isolated, they can be
combined into a transmuting agent. To guide his practice, Jābir now
borrows a concept derived ultimately from Greek medicine, but probably transmitted in more developed form through Arabic physicians.
The physician Galen of Pergamon (AD 129–99) organized Hippocratic
medicine using a system parallel to Aristotle's qualities and elements.
In analogy with the four elements fire, air, earth, and water, the human
body contains four humors: blood, phlegm, black bile, and yellow bile.
These bodily humors were, like the elements, linked to Aristotelian primary qualities: phlegm is cold and wet, black bile is cold and dry, and
so forth (fig. 2.1). When the four humors are in proper balance, or "tem-
perament," the body is healthy. But the quantity of each humor varies

depending on diet, activity, location, season, and other factors; when they fall out of balance, illness results. Consequently, the physician must determine the imbalance and supply a treatment that counteracts it.[28] A patient experiencing congested sinuses, runny nose, and depressed activity clearly suffers from an excess of phlegm, an illness that we commonly call—preserving to this day the doctrine of humors and qualities—a "cold," which many (unconsciously Galenic) mothers still believe is caused simply by exposure to cold and wet, rather than by a microorganism. A cure requires either stimulating the body to restore balance on its own, or applying contraries, that is, hot and dry medicines to restore the humoral balance.

Jābir's transmutational system works in just the same way. He teaches that each metal is composed of a precise mathematical ratio of the qualities. Hot and wet predominate in gold, for example, while in lead, cold and dry predominate. Turning lead into gold therefore involves introducing more hot and wet or reducing the cold and dry.[29] Thus, Jābir devises a practical method for getting down to work. After successfully isolating hot and wet, the alchemist can combine them into a substance that when added to lead should adjust its proportion of qualities to the ratio found in gold, thereby transforming it into the precious metal.

Jābir's terminology for his transmuting agents underscores the link to medicine. Greco-Egyptian alchemists had used the word *xērion* to describe the agent of transmutation, a word that originally referred to a type of medicinal powder used for curing wounds. Jābir uses the same medicinal term, but transliterated into Arabic as *al-iksīr*. (To convert *xērion* into Arabic, remove the Greek grammatical ending *-ion*, and add the Arabic definite article *al-* and then an *i* to aid pronunciation.) This Arabic word for the alchemical transmuting agent has come down to us as *elixir*, a term still used for substances, especially medicines, with marvelous effects. Jābir's elixirs "heal" the metals by adjusting their ratio of qualities, just as a medicine heals the sick by adjusting their ratio of humors. Consequently for Jābir, each metal requires a specific elixir, just as each patient requires a specific medicine. Each elixir is composed of mathematically precise amounts of the isolated qualities that, when added to those already present in a particular metal, sum up to the perfect ratio needed for gold. Simple, logical, and elegant!

Jābir's theory of elixirs is novel and original. Because his elixirs are simply combinations of the four qualities in the right proportions, they can be prepared from virtually anything, because hot, cold, wet, and dry exist in all material substances. This idea stands in stark contrast to the Greek authors who claim that the greatest secret in alchemy is discovering the correct substance from which the transmuting agent is to be made, and who generally stipulate that it is something in the mineral realm. The earliest Jābirian text, the *Book of Mercy*, agrees with the Greek authors, but the later *Seventy Books* prefers to start with animal substances. This change of heart may have stemmed from frustrations in practice: it is easy to decompose animal and vegetable substances by distillation, but difficult or impossible to do so with most mineral substances. Despite the theoretical complexity of the Jābirian texts, and how alien their notions sound in terms of modern chemistry, it is crucial to remember that their authors and readers were actively engaged in practical experimentation. They had extensive experience with a wide range of substances. They watched how these substances reacted to heating and with one another. Accordingly, the Jābirian corpus is full of preparative processes and descriptions of operations and reactions of various kinds.[30]

The Jābirian texts describe three levels of elixirs, distinguished by how well the alchemist has purified the qualities (or "natures," as Jabir calls them) that go into their composition. The purer the qualities, the more powerful the elixir. Lazier alchemists could content themselves with fewer distillations and make elixirs of the first two levels, each of which would work moderately and on only a single metal. The master alchemist, however, would not stop until he had purified the qualities to the utmost degree, because the proper combination of those ultra-pure qualities would produce *al-iksīr al-aʿẓam*, the Greatest Elixir, the Philosophers' Stone itself, able to convert any metal into gold.[31]

These ideas all appear in the *Seventy Books*, an early contribution to the Jābirian corpus, dating to the late 800s. As the corpus developed—that is, as other Shi'ite alchemists joined the project and contributed their own ideas and experiences to it—a new level of complexity emerged. Some readers might already be asking the questions that a later Jābirian alchemist must have asked: If we must add to the qualities present in a base metal, don't we first need to know exactly how much of each is present? How do we know how much hot, cold, wet, and dry exist in lead, so that we can know how much more or less is needed to turn it into gold? Today we would think automatically of some empirical method of analysis involving separation and weighing, and so apparently did the earlier author(s) in the Jābirian corpus. But by the middle of the tenth century, Jābirian authors had begun to think differently about the problem. The starting point continued to be Galenic medical ideas, but the suggested methods for putting them into practice veered off into other—perhaps surprising—areas.

One of Galen's contributions to medicine relates to this very problem of measurement. He introduced a semiquantitative scale to quantify how far off-balance a sick patient's humors were. He subdivided the qualities (hot, cold, wet, dry) into four degrees of intensity, and then classified drugs and illnesses into those degrees. Galen's idea relates to questions of dosage. After all, if a patient was only mildly "cold" (that is, in the first degree), the application of a medicine intensely hot (in the fourth degree) would be dangerous rather than helpful, because it could throw off the balance even further in the opposite direction. The illness and the medicine used to treat it had to be *balanced*.

Jābir's *Books of the Balance* (*Kutub al-Mawāzīn*) apply a modified version of this system to transmutation. Exactly how much more intense is Galen's second degree than the first? Jābir asserts that the relationship between the four degrees is as 1:3:5:8; that is, the second degree is three times more intense than the first, the third five times, and the fourth eight times. Then he subdivides each of these four degrees into seven *grades*, giving a total of twenty-eight levels of intensity for each quality. Next, to ascertain exactly how hot, cold, wet, and dry a particular substance is, he takes a surprising turn, not into quantitative analysis but into Pythagorean number symbolism.

Jābir makes a chart with the four qualities arranged at the top of four columns, and the seven grades of intensity arranged in seven rows, giving a table with twenty-eight boxes. He fills the boxes with the twenty-eight letters of the Arabic alphabet, one in each box, thus assigning a quality and a grade to each letter. Then he takes the name of a substance, say, *usrub* (lead), which in Arabic is written with four letters ('*alif*, *sin*, *ra*', and *ba*'), and analyzes it using the table. The table assigns '*alif* to hot in the highest grade, and since it is the first letter of the word, it is classed as first degree. Thus, we discover that lead is hot in the highest grade of the first degree. The chart assigns the letter *sin* to dry in the fourth grade, and because it is the *second* letter in *usrub*, lead must be dry in the fourth grade of the *second* degree. And so on for the rest of the word. Once this alphabetic analysis is done, another table converts degrees and grades into actual *weights*, allowing the relative weight of each quality present in lead, or any other substance, to be ascertained. Thereafter one can calculate exactly what weight of each quality needs to be added to a given weight of lead in order to bring its composition to the proportions found in gold.

Modern readers should not feel disappointed by what seems to be an arbitrary system rather than something "scientific" in the modern sense. Instead, it provides the opportunity for reflecting on a crucial point for the history of science. People today and people of the past often do not share the same vision or expectations of the world, nor do they necessarily approach the world in the same way. Their questions were not our questions, nor were their ways of answering them necessarily our ways. What seems arbitrary to one expresses a profound law of nature to the other; what seems an insight into the design of the cosmos to one appears as mere trivia to the other. Recognizing these differences helps us avoid the error of projecting our own knowledge and expectations onto the past as measures of its value.

For Jābir, his alphabetical system is *not* arbitrary; it incorporates eternal verities about the way the world is. Consider first the ratio he gives for the four Galenic degrees of intensity, 1:3:5:8. Where does it come from? The four numbers add up to seventeen. For Jābir, seventeen is the fundamental number for the world—his equivalent, if you will, of what the speed of light or Planck's constant means for us. He did not pick this number out of a hat. This number recurs throughout the ancient Mediterranean world, beginning with the Pythagoreans, a secretive group founded in the sixth century BC, for whom mathematics was key not only to the material world but to philosophy, religion, and life. Their central dictum that "the world is number" proved enormously influential in various ways, even to the present day. Numbers form the basis of what is, and numbers have meaning in themselves, detached from what they acquire by being used to count or measure something. Accordingly, the Pythagoreans sought meaning—both physical and especially metaphysical—in numbers and mathematical relationships.[32] According to Pythagorean principles, seventeen is the sum of two important numbers, seven (which expresses divinity) and ten (which expresses completion). It is also the seventh prime number, the sum of the 9:8 ratio that describes the relationship of adjacent notes in the musical scale, and (nearly) the length of the hypotenuse of a right isosceles triangle of height twelve. The number makes an indirect appearance even in the gospels. When the resurrected Christ tells the apostles to cast their nets into the sea, they catch 153 fish, which is the "triangular number" of 17, that is, the sum of the first seventeen integers.[33] (It makes me wish the ancients had known about the North American cicadas that make their noisy appearance by the billions once every seventeen years.) Seventeen is also the number of consonants in the Greek alphabet, and in some Neoplatonic systems the vowels represent the immaterial and the consonants the material. With this background in mind, we can see how Jābir viewed seventeen as a fundamental number for all material substances.

Just as numbers held for premoderns a meaning and significance well beyond their use as quantities, so also did words represent much more than conveniences for human communication. It was not arbitrary or naïve for Jābir to analyze the Arabic names of substances in order to learn something about the substances themselves. Muslims believe that the Qur'an was *dictated* to Muhammad—in contrast to orthodox Christians, for whom the holy scriptures were inspired by God but expressed in words chosen by the sacred writers. God's use of Arabic (transmitted by the archangel Gabriel) means that Arabic is a divine language. As such, Arabic words are not arbitrary signifiers of things. They are instead God's names for the things He created, and therefore carry profound meaning and true links to the objects they name. Analyzing the name of a thing can thus reveal something about the thing itself. The same thinking forms the basis of that branch of Jewish Kabbalah known as gematria, and its Christian versions that were explored in the Middle Ages and Renaissance.

Once this difference of worldview is understood, we could even argue that Jābir's underlying aspirations are actually quite similar to our own. His fundamental goal was to classify and quantify natural substances mathematically so that practitioners could work with them in precise, quantitative ways. Seen from this perspective and in context, the system is actually an advanced attempt to standardize and understand mathematically what he saw as the intrinsic qualities of substances. Jābir sought to grasp, unify, and work with the underlying rules and phenomena hidden behind what is visible in the natural world, a fundamental feature of virtually all scientific fields today. Moreover, the successive elaborations of the "Jābirian" method might well have resulted from failed empirical attempts to get earlier theoretical conceptions to translate successfully into practice.

This final version of Jābirian alchemical theory was not taken up by later alchemists. It was perhaps too complex, and it did not translate out of Arabic. The simpler Mercury-Sulfur theory was, however, widely adopted, and the Latin West learned about it from Jābir, from later Arabic authors who followed him, and even directly from the *Book of the Secret of Creation*. If it seems that the Mercury-Sulfur theory and the four-element theory of composition exist in an uneasy or unclear relationship to each other, this is in part due to the evolution of ideas within the Jābirian corpus. Still, it might be proposed (as some alchemists in fact did) that Mercury is the carrier of cold and wet qualities, while Sulfur carries hot and dry, or that the elements combine to produce Sulfur and Mercury, and these go on in turn to produce the metals.

Alchemical Secrecy and Literary Style

The Jābirian corpus also carries stylistic features that left their mark on subsequent alchemical writers. The first of these is the dispersion of knowledge (*tabdīd al-'ilm*), a method ostensibly for helping to preserve secrecy. Jābir states that "my method is to present knowledge by cutting it up and dispersing it into many places."[34] The idea is that the entirety of Jābir's teaching cannot be found altogether in one place; instead, he distributes a single idea or process piecemeal through one or several books. This technique partly fulfills the charge given to Jābir by his supposed master, Ja'far: "O Jābir, reveal the knowledge as you desire, but such that none have access to it but those who are truly worthy of it."[35] Kraus suggested that the practical reason for this dispersion was to hide the multiple authorship of the Jābirian corpus, allowing later authors to claim that earlier texts were "incomplete," thereby making room for new additions to the corpus, binding its various layers together as a whole, and explaining away contradictions between books.[36] Whatever the original cause, this method would be imitated in many later alchemical texts, and accordingly, Latin alchemists often quoted the motto *Liber librum aperit*, that is, "One book opens another."

The Jābirian corpus shows an increased level of secrecy over earlier texts, but there is little use of *Decknamen* (although these are common in other Arabic alchemical texts) or the kind of enigmatic allegory deployed by Zosimos.[37] Yet the Jābirian writers are clearly aware of these techniques. Indeed, Jābir exclaims with his characteristic humility, "I have revealed the whole of the science without using enigmas in the least letter; the only enigma consists in the dispersion of knowledge. By God, no one in the world is more generous nor has more mercy on the world and its inhabitants than I!"[38] The reader may be excused for thinking this statement rather disingenuous.

Another stylistic feature that became characteristic of alchemical writing is an "initiatic style."[39] That is, the author writes in a self-consciously grand manner, speaking as the master of a closed circle and addressing his readers as postulants. This initiatic style is evident in portions of the Jābirian corpus where it arises partly from casting the works as teachings of Imam Ja'far, and partly—like the enhanced secrecy—from characteristics of contemporaneous Isma'ili groups. These groups adapted a secretive, initiatic nature from the Neoplatonic philosophies they followed and adopted it as expedient policy, since the more "ultra-Shi'ite" factions were considered religiously unorthodox from the point of view of majority Islam. Yet influences local to the religious and political context in which the Jābirian corpus was written were propagated throughout the rest of alchemy as later writers strove to imitate Jābirian writings. Indeed, Robert Boyle (1627–1691), exasperated at trying to decipher later alchemical texts, would burst out, "These writers, after they have frequently called their reader their son and made solemn professions . . . that they will disclose to him their secrets . . . put him off with riddles instead of instructions."[40]

The Turba *and al-Rāzī's* Secret of Secrets

Around the year 900, another alchemical classic first appeared, known generally by its Latin title *Turba philosophorum* (*The Throng of Philosophers*). The work is cast as a meeting of Greek philosophers. Nine pre-Socratics are named, such as Empedocles, Anaxagoras, Leucippus, and others, and Pythagoras presides over the assembly. Together these characters debate the composition of matter and cosmology, each providing a version of the ideas (sometimes rightly, sometimes wrongly) attributed to their pre-Socratic namesakes. The anonymous Arabic author seems to have drawn upon an early third-century book against heresies by the church father Hippolytus, and upon the writings of Olympiodorus that compare earlier Greek philosophers with later Greco-Egyptian alchemists, but all this Greek material is translated into an Islamic context. Much of the point of the *Turba* is to demonstrate that the God of Islam is the Creator, that the world is of a uniform nature (monism again), and that all creatures are composed of the four elements.[41] Clearly, this work is of a very different nature from the Jabīrian corpus—it contains no practical instructions and nothing explicitly about chrysopoeia. Nevertheless, many later alchemists esteemed it for its discussion of the nature of matter, a topic obviously of central importance to alchemy. The *Turba* also further indicates the important role of Greek philosophical ideas, and their continued development, within the Islamic world.

Abū Bakr Muhammed ibn-Zakarīyya' al-Rāzī (circa 865–923/4), often known in the Latin world as Rhazes, exemplifies a very different sort of Arabic alchemist. He was born in the city of Rayy in Persia, and became one of most famous physicians and alchemical writers in the Islamic world. His works remained authoritative texts in Europe until the 1600s. It is recorded that al-Rāzī wrote at least twenty-one books about alchemy.[42] He rejected Jābir's theory of balance, but adopted the Mercury-Sulfur theory of the metals, and added to it the notion that sometimes a salt is contained in the metals as well. His best-known work, the *Book of Secrets* (*Kitāb al-asrār*), also called the *Book of the Secret of Secrets* (*Kitāb sirr al-asrār*), was written for one of his students.[43] Often reading almost like a laboratory manual, it begins with a systematic classification of naturally occurring substances—volatile substances ("spirits"), metals, stones, vitriols, borax, and salts—and the different varieties of each. Al-Rāzī carefully describes how to recognize and purify each of them, and continues with descriptions of the apparatus and furnaces needed for various operations. Techniques such as distillation and sublimation are described next. Dozens of recipes then follow for a wide variety of products. The careful detail with which they are given indicates that they are the product of considerable practical experience. The richness of al-Rāzī's inventory of substances and apparatus reveals how substantially Arabic alchemists had expanded the material and technical content of alchemy beyond what earlier Greek writers had known.

Al-Rāzī was also clearly interested in transmutational endeavors—many of the recipes in the *Book of Secrets* yield products supposedly leading to transmutations of one sort or another. Moreover, he added a new dimension to alchemy's goals, namely, the turning of stones, rock crystal, and even glass into precious stones. These transformations, like those of the metals, were to be carried out by means of specially prepared elixirs. The *Book of Secrets* ends with recipes for a variety of elixirs made from mineral and organic substances such as eggs and hair. Yet much of the content of the book does not relate immediately to transmutation. Alchemy (or, for al-Rāzī, *al-kīmiyā'*) covers much more than just chryso-poeia. The restriction of the word *alchemy* to the context of making gold is a development dating many centuries after al-Rāzī's time. In fact, that narrow definition which now seems so natural did not emerge until the end of the seventeenth century. Before that time, *alchemy* referred to all the processes and concepts we might today consider broadly "chemi-cal." In other words, al-Rāzī's classification system of substances is most certainly a central part of the history of alchemy, even when it does not relate to transmutation.

Ibn-Sīnā and the Critics of Transmutation

As alchemy expanded and developed during the Arabic period, so did reactions against it in the form of criticism, skepticism, and the denial of alchemical claims. If such anti-alchemical literature existed in the Greco-Egyptian period, it no longer survives.[44] But in the Arabic world, dissent became common. Al-Kindī (died 870), a prolific writer deeply interested in Greek philosophy and scientific thought, wrote a short treatise against the reality of chrysopoeia, although it is now lost.[45] Al-Rāzī, on the other hand, came to transmutation's defense and wrote his own tract—also now lost—refuting al-Kindī.[46]

The most influential attack on chrysopoeia came from the pen of ibn-Sīnā (circa 980–1037), generally known as Avicenna in the Latin world. Like al-Rāzī, ibn-Sīnā was a Persian, and wrote medical texts, most no-tably his authoritative treatise *The Canon* (*al-Qānūn*), which became a fundamental authority for European medical schools until the sev-enteenth century. Yet he also addressed the topic of alchemy. One of ibn-Sīnā's works, the *Treatise on the Elixir* (*Risālat al-iksīr*), claims wide familiarity with texts both for and against alchemy (he was impressed by neither), and is cautiously positive about chrysopoeia. Yet the attribu-tion of the *Treatise* remains a topic of debate; if it really is ibn-Sīnā's, it may express early ideas.[47] What is clearer is that his better-known *Book of the Remedy (Kitāb al-shifā')* comes to a different conclusion. This unquestionably authentic work contains a section about minerals where ibn-Sīnā discusses the formation of minerals and metals, adopting the Mercury-Sulfur theory that by his time had become standard. But unlike al-Rāzī and other alchemical writers, ibn-Sīnā then goes on to deny the possibility of metallic transmutation: "As to the claims of the alchemists, it must be clearly understood that it is not in their power to bring about any true change of species."[48] The core of ibn-Sīnā's rejection involves two closely related points: human weakness and human ignorance. For the first, he states that the power of human industry is simply weaker than nature: "Alchemy falls short of nature . . . and cannot overtake her."[49] Or, as he states in another book, "Whatever God created through natural powers cannot be imitated artificially; human industry is not the same as what nature does."[50]

Ibn-Sīnā believes that artificially prepared things can never be identical to natural ones, whether we are talking about gold, gems, or anything else. Thus, he would agree with those people today who (incorrectly) believe that the vitamin C in an orange is somehow different from that which is produced chemically for vitamin supplements. Regarding human ignorance, ibn-Sīnā claims that what we sense and identify as the differences between metals—that is, what alchemists endeavor to alter—are not their true, essential differences, merely superficial ones. The true differences are unknown to us, hidden within the very essences of things. If we do not know what those true differences are, we cannot produce or change them correctly. Therefore, given this combination of weakness and ignorance, alchemists attempting to transmute metals "can make excellent imitations . . . yet in these [imitations] the essential nature remains unchanged; they are merely so dominated by induced qualities that errors may be made concerning them."[51] In other words, alchemical gold might very well look like gold, have all the apparent characteristics of gold, and convince at least some people that it is indeed gold, but it is not really *true* gold.

Ibn-Sīnā's denial of the possibility of true transmutation turned out to be extremely influential, for this section of his *Book of the Remedy* would later be translated into Latin and widely circulated in Europe, often under the weighty name of Aristotle himself (see chapter 3). But while ibn-Sīnā's critique provided ammunition for those who sought to discredit alchemy, it did little to dampen the interest of those who pursued it. Several later Islamic alchemists wrote refutations of ibn-Sīnā—most notably al-Ṭughrā'ī in the early twelfth century.[52] There are two important points to stress. First, after their emergence in the Arabic world, critical views of alchemy never vanished; alchemy would remain forever after a controversial subject, with parties vigorously arguing for or against it down through the centuries. Second, while ibn-Sīnā's critique is based on philosophical principles—influenced in part by Aristotelian thinking—his concession that alchemists can make something that looks so much like gold that it can fool people leads naturally to another sort of criticism: the connection of transmutation to intentional fraud.

Tales of the charlatan alchemist are not uncommon in the Arabic world, though the earlier Greek world shows almost no sign of such stories.[53] Al-Kindī's lost anti-chrysopoetic treatise reportedly cataloged tricks used by such charlatans to deceive the unwary. But 'Abd al-Raḥmān al-Jawbarī provides a longer inventory of shady alchemical dealings. Around the year 1220, al-Jawbari wrote a book called *The Revelation of Secrets* that details a variety of cheats and swindles. He recounts the sleights of hand used by false alchemists to dupe the unwary—gold hidden inside charcoals, under the false bottom of a crucible, or within metal implements that is made to appear at the right moment as if produced by transmutation. Curiously, alchemists would be accused of many of the very same tricks into the eighteenth century. Among his many anecdotes, al-Jawbari includes one about a man who asks a goldsmith to sell ingots of silver for him, and then befriends the goldsmith with his generosity. When the man's evident wealth disappears, the goldsmith asks the cause, and learns his new friend is an alchemist who has run out of the transmuting elixir, and now—through various misfortunes—has neither place nor resources to produce more. The goldsmith (of course) invites him into his house and provides him with the necessary equipment and materials, including a substantial amount of gold and silver. The alchemist sets to work making a fresh batch of elixir with the promise to share it. Needing a particular mineral to complete his work, the alchemist sends the goldsmith off to collect it. When the goldsmith returns home, the "alchemist" is gone, along with the gold and silver.[54]

Al-Jawbari's purpose was to entertain the reader with amusing tales of the cleverness of con men and the gullibility of their victims. The proportion of fact to fiction in these anecdotes is impossible to assess, and so it remains unclear whether such traveling impostors really existed or if these accounts were simply humorous and plausible fictions. Nevertheless, such stories provide a tantalizing look at one role the figure of the alchemist played in Islamic popular culture. Unfortunately, we have precious few surviving sources—or at least few that are currently known and available—that can tell us much more about this important point in the history of alchemy. We will have to wait until we reach early modern Europe—where sources are more plentiful—to explore that particular topic.

Another insight into the actual lives of alchemists in the Islamic world occurs much later, in a sixteenth-century work by Leo Africanus, a freed slave and Christian convert sent by Pope Leo X to compile a descriptive account of Northern Africa. Leo gives a highly unflattering account of the many alchemists who inhabited the Moroccan city of Fez. They stink of sulfur, and assemble nightly at the chief mosque to debate their processes. Some among them seek the elixir using the works of Jābir, while others seek ways to extend precious metals. "But their chiefest drift is to coin counterfeit money, for which cause you shall see most of them in Fez with their hands cut off."[55] (Leo does not explain exactly how they could persist in practicing alchemy without hands.) Charges of forgery and counterfeiting would continue to vex alchemists in both Muslim and Christian worlds.

Alchemy continued to flourish in the Arabic world long after al-Rāzī and ibn-Sīnā.[56] The historian of alchemy E. J. Holmyard was taken to see a working subterranean alchemical laboratory outside Fez in the 1950s.[57] (Such places continue to exist in Europe and North America as well.) Moreover, I have heard anecdotally from colleagues of their meeting Muslim alchemists still at work on transmutation even today in Egypt and Iran. But having now sampled the theoretical and material sophistication that alchemy acquired in the Islamic world, it is time to move on to alchemy's third cultural context. By the twelfth century, the Dār al-Islām, or Abode of Islam, shared borders with another civilization in three places—Palestine, Sicily, and Spain—and that other culture, Western Christianity, had begun a process of vigorous growth and renewal. Latin Europe was ready to discover, within the vast intellectual wealth of Islam, the golden promises of al-kīmiyā'.

THREE

MATURITY

Medieval Latin *Alchemia*

W hile the origins of Greco-Egyptian *chemeia* and the beginnings of Arabic *al-kīmiyā'* remain cloudy, the entry of alchemy into the European Middle Ages is not—or at least seems not to be. Alchemy, we are told, arrived in Latin Europe on a Friday, the eleventh of February, 1144. That was the day that Robert of Chester, an English monk at work in Spain, completed his translation from Arabic of a book often given the title *De compositione alchemiae* (*On the Composition of Alchemy*). In his prologue, Robert explains that he chose to translate an alchemical text "because our Latin world does not yet know what alchemy is and what its composition is."[1] That situation would soon change, for Robert's Latin world quickly got to know alchemy very well. Transplanted into its third cultural context, alchemy flourished in Europe for nearly six hundred years. It deeply tinctured European culture and thought and made substantial contributions to the foundations of modern science as we know it.

Robert of Chester's translating work was not done in a vacuum. He was living in one of the most intellectually active and exciting periods of

European history—an era often called the "Renaissance of the Twelfth Century."² All across Europe, new ideas emerged and flourished. The building of the great cathedrals in a new architectural style—later to be (dismissively) called Gothic—had begun. Reforms in law and advances in agriculture joined new styles of literature and music. Under the protective aegis of the church, cathedral schools prospered and a new institution that would change all of intellectual history began to emerge: the university.

Europe was expanding not only its intellectual and artistic frontiers but its geographic ones as well. To the east, west, and south, Christian Europe had begun to push back against Muslim advances made over three hundred years earlier. Once in closer contact with Islamic civilization, especially in Spain (where the Iberian Peninsula was divided between Christian and Muslim rule), Latin Europeans were awed and undoubtedly a little ashamed by what they found. Among their many discoveries were libraries holding volume after volume of books by revered ancients—such as Aristotle, Galen, and Ptolemy—of which they had previously possessed only fragments or digests. Upon those ancient foundations, Muslim scholars had made their own considerable advances, providing Europeans with a wealth of additional knowledge and ideas in astronomy, medicine, mathematics, physics, mechanics, botany, engineering, and fields completely new to Europe—such as *al-kīmiyā'*. In the twelfth century, Europe not only accepted these new ideas but hungered for them. Scholars trekked west across the Pyrenees to Spain, or south to Sicily, or (much less frequently) east to the crusaders' recently founded Latin Kingdom of Jerusalem to learn Arabic and to translate, sending Arabic knowledge as well as recovered ancient Greek knowledge back to the Latin world as fast as they could. Robert of Chester and his companion Herman of Carinthia (also known as Herman the Dalmatian) were among the translators who journeyed to Spain.

Amusingly enough, it is Morienus the monk who reappears as the flag bearer of alchemy's transmission to Latin Europe. Robert's *De compositione alchemiae* is a translation of the supposed instructions of Morienus (aka Marianos) to Khālid ibn-Yazīd for making the Philosophers' Stone. Here Robert uses his newly coined Latin word *alchemia* (which he calls "unknown and surprising") to mean not the whole subject but rather just the Philosophers' Stone itself—"a material substance compounded out of one thing . . . naturally converting substances into better ones."³

Other translations of Arabic alchemical works followed soon after—Jābir, Balīnūs, al-Rāzī, and ibn-Sīnā, among others—and the word would quickly come to refer to the entire discipline, as its cognates had done in Greek and Arabic.[4]

The Recipe Literature in Europe

Although a developed science of alchemy was new to Europe, metallurgical and productive processes were already well established there. European craftsmen and artisans possessed a range of practical know-how for the production of a variety of substances—alloys, pigments, dyes, techniques for metalworking, and so forth. Several early medieval manuscripts record this knowledge. They perpetuate the ancient recipe literature tradition to which the Greco-Egyptian Stockholm and Leiden Papyri and the *Physika kai mystika* of pseudo-Democritus belong. In fact, an Italian text called *Compositiones variae* (*Various Compositions*) and probably dating from around 800 actually contains a word-for-word Latin version of one of the recipes recorded in the Leiden Papyrus. This compilation, as well as the slightly later and more extensive *Mappae clavicula* (*Little Key to the World*), indicates how workshop recipes and practices were handed down for centuries.

While these written texts testify to a transmission of knowledge—in these cases mostly through Byzantium—the texts themselves are predominantly literary compositions. That is to say, the *Compositiones* and *Mappae* were not handbooks for craftsmen; an artisan would not have kept one of them in his workshop as a handy reference guide. These texts were compiled from a variety of sources by scribes who scarcely if ever set foot in a medieval *laboratorium* (workshop), and almost certainly never applied their ink-stained fingers to the artisanal crafts described.[5] Thus, they contain recipes of widely diverse date and origin, quite a few of them garbled by the scribes' unfamiliarity with the techniques and terminology.

One exception to this generalization is the most famous of craft books: *On Diverse Arts* (*De diversis artibus*), written about 1125 by a monk calling himself Theophilus. It describes a wide variety of substances and technical details useful to monastic craftsmen for making pigments, glass, cast metal objects, and alloys.[6] Most of its recipes are so clearly described that they can be replicated fairly readily today. This implies that Theophilus himself had firsthand knowledge of the operations and processes he describes. Yet amid his otherwise perfectly straightforward processes, there exists a curious recipe that may mark one early percolation of Arabic alchemy into Latin Europe before Robert of Chester's translating activity. Amid descriptions of various kinds of gold, Theophilus includes a recipe for "Spanish gold . . . compounded from red copper, basilisk powder, human blood, and vinegar."[7] Copper, vinegar, and human blood are easy to obtain (although the last perhaps rather unpleasantly), but basilisk powder was probably not readily available in the average monastic workshop, not even in the back of the cupboard. Readers who know their bestiaries (or their *Harry Potter*) will recognize the basilisk as a hideously deadly reptile able to kill with its glance alone. But Theophilus explains that the "gentiles" (that is, Muslims) have commendable skill in making basilisks. They lock up two old roosters in a narrow place and overfeed them until they copulate and lay eggs. The eggs are given to toads, who hatch them into chicks that soon grow serpents' tails and mature into basilisks. The basilisks are raised underground in kettles and later incinerated, their ashes mixed with vinegar and blood, and the paste smeared onto plates of copper. Exposure to fire then turns this copper into fine gold.

What we may be seeing here is a garbled alchemical allegory being taken literally. Perhaps Theophilus included the process on account of its exotic nature—it's doubtful that either he or any of his readers were tempted to try it for themselves. Remarkably, historians of science have recently found a similar recipe for basilisks and the use of their ashes for making gold—possibly translated from some part of the Jābirian corpus—in a Sicilian manuscript, and have even outlined a plausible transmission route between that manuscript and the author of *On Diverse Arts*.[8]

The Emergence of Latin Alchemy and "Geber"

The translation of Arabic alchemical works dwindled to a trickle within a hundred years, around the middle of the thirteenth century. By that time, Latin authors had begun writing their own books on *alchemia*.[9] Centuries earlier, when alchemy was appropriated from the Byzantine world by the Arabs, the first original Arabic compositions appeared under Greek pseudonyms. Now in Europe, an analogous situation prevailed: many of the earliest Latin authors wrote their books under *Arabic* pseudonyms. In both cases, pseudonymous authorship was intended to give books greater authority by making them seem older, more venerable, and part of a culture recognized as more advanced. To add to the sense of déjà vu, the most influential of these thirteenth-century Latin alchemical compositions appeared under a very familiar name, that of Jābir, rendered in medieval Latin spelling as Geber. Thus, the "Jābir problem" discussed in the previous chapter had yet another dimension: whether the Latin books known under the name of Geber were translations of Jābir, or whether they were native Latin productions. Historians of science argued vociferously over whether Geber was really Jābir. Recent scholarship has settled the issue: he was not. Geber was a late thirteenth-century Latin author.

Many writers continue to confuse the two today, and a stubborn few continue to battle in favor of an Arabic identity for Geber. Geber himself did not make settling the issue easy. He cites no sources by name which might allow us to date or place him. He adopts the initiatic style typical of the Jābirian corpus (but only at the very beginning and end of his book) and rewrites sections of Jābir's *Seventy Books* to include in his own treatise. He even peppers his text with what seem to be characteristically Arabic grammatical constructions and expressions translated into Latin.

The author concealed behind the pseudonym of Geber is probably an Italian Franciscan friar and lecturer named Paul of Taranto.[10] Paul wrote a nearly contemporaneous alchemical text that bears a striking similarity in style and content to "Geber"'s. Although Paul's writings draw heavily from Arabic sources, particularly from Jābir and from al-Rāzī's *Book of Secrets* (*Kitāb al-asrār*, translated into Latin as *Liber secretorum*), they also show striking originality and a detailed familiarity with practical alchemical processes. His *Theory and Practice* (*Theorica et practica*) classifies mineral and chemical substances much like al-Rāzī did, but shows more interest in describing and categorizing substances based on their observable chemical and physical properties. Undeniably, Paul's works display the results of extensive practical testing and experimentation, and a level of rigor and theoretical synthesis rarely found in Arabic sources. This difference may come from the Christian West's taking more seriously Aristotle's charge to discover the true natural causes of things than was widely the case in the Islamic world. Typical of Paul's work is a desire to harmonize theory and practice at a deep level by developing a rigorous physical underpinning that could explain observed phenomena coherently.

Such ideas are more fully expressed in the *Summa perfectionis* (*The Sum of Perfection*). In the Middle Ages, the Latin word *summa* commonly referred to an exhaustive "textbook" of one or more subjects, as in St. Thomas Aquinas's *Summa theologica*. Accordingly, the *Summa perfectionis* is a comprehensive text about alchemy. It begins with arguments for and against the possibility of chrysopoeia (deciding in favor of it), and continues with a detailed summary of the state of knowledge about metals and minerals, including methods for purifying and working with them. Thereafter follow sections addressing practical operations and apparatus, while the last section provides a fascinating examination of the nature and properties of the metals and an account of the different grades of transmuting agents. The book concludes with an account of assaying, that is, how to determine the purity of precious metals—a necessary skill if the alchemist is to test the quality of the gold and silver he hopes to produce. The *Summa* became one of the most influential alchemical books of the Middle Ages and remained an authoritative text until the seventeenth century.

According to Geber, the alchemist may carry out his art using "medicines" (by which he simply means chemical agents) having three degrees of strength. The least potent changes only the superficial appearance of the base metal, making it merely look like gold or silver. Geber provides several practical tests using fire and corrosives to demonstrate that true transmutation has not occurred. Only the most potent medicine, that of the "third order," can truly transmute, and it exists in two forms—one for making silver and the other for making gold. This threefold division is one thing Geber borrows from Jābir.[11] What he does not adopt, however, is Jābir's idea that animal and vegetable substances could be used to make transmuting elixirs—for Geber, as most European alchemists would eventually agree, the Philosophers' Stone must be made from mineral substances only.[12]

More strikingly, the *Summa* incorporates a coherent matter theory that explains laboratory observations and undergirds methods of transmutation. This theory is based on two earlier ideas: the Arabic Mercury-Sulfur theory of the metals, and an idea out of Aristotle. Although the Greek philosopher explicitly denied the existence of indivisible atoms, in two places in his writings he made comments that could be interpreted to suggest, or at least support, a theory of matter based in the existence of tiny particles of some sort. In one passage, he claims that there was a lower limit of size that any piece of a substance could have and still maintain its identity. A lump of gold, repeatedly divided in half, would eventually become so small that one further cut would no longer result in two smaller pieces of gold; the particle would have become too small to support the properties of gold. These tiniest pieces of a substance came to be called in Latin *minima naturalia* (the smallest natural things). But more important for Geber—and in fact for all alchemists—was the fourth book of the *Meteors*, where Aristotle (or perhaps a follower of his) consistently invokes the idea of "parts" (*onkoi*) and "pores" (*poroi*) that exist within seemingly solid substances. These parts and pores are used to explain a wide range of observations, phenomena, and physical properties.[13]

Geber draws on these ideas, especially the latter, and combines them with the Mercury-Sulfur theory. According to the *Summa*, the metals are produced from the coalescence of minute "parts" of the metallic principles Mercury and Sulfur. In different metals, these minute parts are of different sizes, and in the case of base metals, they are intermingled with earthy particles. Although this system bears some resemblance to an atomic theory, Geber's system is not truly atomic, because the "tiniest parts" (*minimae partes*) he describes are neither indivisible nor permanent.

But Geber does use this system to explain an array of physical properties and chemical changes. Take for example the fact that a piece of gold is far heavier than an equally sized piece of tin; in modern terms, gold has a much higher density. Geber explains this observation based upon the way the constituent particles of Mercury and Sulfur are packed. In gold, they are very small and packed as tightly together as they can be—in what he calls "the strongest putting-together" (*fortissima compositio*)— while in tin they are larger and poorly packed. A piece of gold thus contains more matter, leaving less space between the constituent parts than in an equal volume of tin; thus, the gold weighs more.[14] Geber explains gold's stability with the same theory: because its constituent parts are so tiny and so tightly packed together, no pores or crannies are left open whereby fire or corrosives can attack and penetrate the metal to break it apart. A base metal such as lead, however, is badly "put together" (*compositum*), so when it is roasted on the fire, the fire can enter the pores of the metal and break it down into a powder. (What Geber describes here is the process of oxidizing lead into powdery lead oxide by means of roasting.)

The same theory also explains chemical operations. Sublimation—the purification of a volatile material by converting it from a solid into a vapor and then recondensing the vapor into a solid—occurs with substances whose particles are not held tightly together. The heat of the fire causes the tiny parts to dissociate from one another; the smallest particles, which Geber considers the purest, rise up as a fume, away from the larger and heavier ones, which are left behind unsublimed as dross at the bottom of the vessel.[15] Although only some subsequent alchemists followed Geber's ideas, his theory would persist and develop as one important thread among the various theoretical traditions of European alchemy.

Alchemy Becomes Controversial

A modern reader of the *Summa* might not realize that the book was written amid a century's worth of intense controversy over the promises and aims of alchemy. At the center of this controversy was that polemical text written against chrysopoeia by ibn-Sīnā almost two centuries earlier and half a world away. Around 1200, an English translator named Alfred of Sareshal rendered that section of ibn-Sīnā's mineralogical text into Latin as *De congelatione et conglutinatione lapidum* (*On the Congelation and Gluing-Together of Stones*). Alfred's translation of this short text ended up being placed in manuscripts at the end of a translation of Aristotle's *Meteors*; pairing the two works made sense, since both deal with the origin of minerals. But, perhaps due to a hasty copyist who did not separate the two texts clearly, many readers thought that ibn-Sīnā's words were part of Aristotle's text. Given the enormous esteem that Aristotle had acquired by the thirteenth century, this mistake greatly enhanced the power of ibn-Sīnā's ideas—for both good and ill. On the one hand, the first part of ibn-Sīnā's text helped establish the Mercury-Sulfur theory of metals firmly in Latin Europe. On the other hand, the closing section presented a powerful rebuff to aspiring transmuters. Thus, Latin Europe heard the declaration that "art is weaker than nature and cannot follow her no matter how much it tries; let the practitioners of alchemy know that the species of metals cannot be transmuted," in what seemed to be the authoritative voice of Aristotle himself.[16]

Responses came swiftly. An early thirteenth-century work titled the *Book of Hermes* refuted *De congelatione* point by point using both logical analysis and practical experience. Its author pointed out that alchemists could in fact make some substances (like salts) that are identical to naturally occurring ones.[17] St. Albert the Great (circa 1200–1280), known as the Universal Doctor because of his broad learning and wide influence, likewise dissented when he wrote his own study of minerals.[18] Albert's most famous student, St. Thomas Aquinas (circa 1225–1274), took *De congelatione* a little more to heart. St. Thomas echoed its words in saying that alchemists can produce only the *appearance* of natural things; their gold is not true gold, nor are the other things they produce the same as natural products, even if they display all the same properties. Nevertheless, St. Thomas elsewhere allows that *if* alchemists are able to harness the powers of nature and use them to produce gold in the same way that nature does, then that gold would be true gold, and could be legitimately sold and used as such.[19] The critical factor lies in the precise method alchemists use—but could alchemists really identify and utilize nature's own means? One of St. Thomas's followers, Giles of Rome (circa 1243–1316), went further. He recognized that *De congelatione* was not Aristotle's but ibn-Sīnā's (as Albert himself had suspected), yet he still used its arguments to declare that no matter how many tests alchemical gold passes, and even if no sensible difference between it and natural gold can be detected, it is still not the same thing as gold from the earth. If alchemy can make gold, he concludes, "it ought not be used as money, because gold and such metals are sometimes used in medicines and other things serviceable to the human body. If therefore such gold were alchemical, it might greatly harm the human complexion."[20]

Some contemporaneous alchemical authors agreed that artificially produced metals show subtle differences from the natural ones. The *Little Book of Alchemy* (*Libellus de alchimia*) attributed to St. Albert the Great claims intriguingly that alchemical metals are "equivalent to all natural ones for all purposes," with the odd exceptions that alchemical iron is not attracted by magnets; alchemical gold lacks medicinal properties; and wounds made by alchemical gold fester while those made by natural gold do not. In the *Book of Minerals*, St. Albert reports, "I have had tests made on some alchemical gold, and likewise silver, that came into my possession, and it endured six or seven firings, but then, all at once, on further firing, it was consumed and lost and reduced into a sort of dross."[21] One wonders what sort of a substance the Universal Doctor had acquired, and from whom!

Giles of Rome reveals a concern that should sound familiar to modern readers. He is worried about hidden properties and unknown effects. Even if alchemically produced gold conforms in every way to natural gold's properties—in color, density, softness, resistance to corrosion, and so forth—there still might be *something* we do not know about, *something* we cannot foresee, do not think to look for, or cannot detect. Giles bases his thinking on the axiom enunciated by ibn-Sīnā that "art is weaker than nature": the creative and productive activity of human beings simply *cannot* reproduce natural things. Nothing artificially produced can equal what is naturally produced. Such thinking remains very much alive today in the notion that synthetic diamonds are not "real" diamonds, and in concerns that bioengineered crops might have hidden noxious properties. Thus, some issues raised about medieval alchemy regarding technology and the relation between the natural and the artificial remain unresolved today.[22]

The partisans of alchemy did not take these attacks lying down. Indeed, it has been argued that their heated defense of the Noble Art and its power to imitate nature represents the first sustained salvo in favor of the power of human ingenuity and technology. One of the loudest voices raised on behalf of alchemy was that of the Franciscan friar Roger Bacon (circa 1214–1294). In 1266–67, he wrote three books at the request of a friend who, in the meantime, had become Pope Clement IV. These writings contain vigorous arguments favoring the study of languages, mathematics, natural philosophy, and alchemy as a means to reform knowledge and strengthen Christendom. In terms of alchemy, Bacon does not merely oppose the notion that art is weaker than nature—he turns it on its head. Human artifice is *not* weaker than nature; it is *stronger*. Alchemically produced gold is *better* than natural gold. Bacon asserts that the same is true of all laboratory products when properly made. Human copies of naturally occurring substances can be made superior to what nature provides.[23] This kind of thinking also endures today; it undergirds modern chemistry. Organic chemists work diligently (and successfully) toward the synthetic production of naturally occurring substances by quicker, more efficient means or with slight structural changes that make them better pharmaceuticals by increasing their medicinal effects or decreasing their toxicity.

Arguments about the power of alchemy to produce the equivalent of natural substances—gold or otherwise—eventually rose to the highest levels of power. It is reported that Pope John XXII convened a debate where the two sides squared off against each other.[24] We have no direct testimony of what happened at this debate, but if such a forum really took place, the papal decretal John XXII issued in 1317 suggests that the pro-alchemy party must not have defended its cause very well, for it begins thus:

> The impoverished alchemists promise riches that they do not deliver, and they who think themselves wise fall into the ditch they have dug. For indeed the professors of this art of alchemy make fools of one another.[25]

The decretal goes on to say that when the alchemists repeatedly fail in making gold, "in the end they feign true gold and silver with a false transmutation," since the possibility of a real transmutation into gold and silver "does not exist in the nature of things."[26] Then they counterfeit coins and pass them off on honest people. In punishment, the decree stipulates, anyone selling or using alchemical metal as if it were natural gold or silver is sentenced to surrender an equal weight of true gold or silver to the public treasury for distribution to the poor.

Although the pope apparently was not convinced that true gold could be made artificially, his statement was not so much a condemnation of alchemy per se as of counterfeiting. It contains no theoretical or practical arguments against chrysopoeia, only concerns about coinage and fraud. Unfortunately for the alchemists, their art was rarely far from such criminal practices in the public mind. The kings of France and England issued similar regulations banning the practice of transmutational alchemy, as did the ruling council of the Republic of Venice.[27] In all these cases, the fundamental concern was to maintain the purity and value of the precious metals upon which economies rested—the same concern that may have motivated Diocletian a thousand years earlier to command that the books of the Egyptians be burned. In fact, whether chrysopoeia could make real gold or only a good imitation thereof, it remained a perilous practice in terms of economic and political stability. If the gold was false, it would debase and pollute the gold supply, and if true, it would lower the value of gold by increasing the amount available. Similar thoughts led the Arab historian ibn-Khaldūn in 1376 to argue against the possibility of chrysopoeia on the grounds that if true, it would foil God's plan for maintaining economic stability in the world—namely, His divinely wise choice to create only a limited quantity of gold and silver.[28] Lawyers continued to debate the licitness of alchemy and the legality of its products from the Middle Ages until the eighteenth century.[29]

As with most papal proclamations then as now, John XXII's decree was largely ignored. Alchemists, including many in Holy Orders, continued to work and write. In England, the allure of a domestic supply of gold was too powerful. Henry IV's 1404 statute against gold making was soon modified, in a very English way, by the awarding of licenses from the Crown to practice alchemy, on the condition that the precious metals produced were to be sold directly to the Royal Mint.[30]

Alchemists themselves reacted to the climate of dissension and concern that came to a climax in the fourteenth century. Yet tracing precise lines of influence is difficult, and modern scholars are still striving to obtain a better understanding of fourteenth-century alchemy. Nevertheless, several changes are evident, and two of them can be reasonably (if not yet rigorously) traced to the atmosphere of greater criticism: enhanced secrecy and the construction of linkages between alchemy and Christian theology.

Secrecy and Theology in Latin Alchemy

As far back as Zosimos, alchemical texts carried injunctions to secrecy and deployed various methods of preserving it, such as the use of *Decknamen* and their extension into allegories. Originating in the protection of trade secrets, this tendency was enhanced in the Jābirian writings by alchemy's connection to a secretive and esoteric Shi'ite sect, and by the addition of the dispersion of knowledge technique, which may have served primarily to cloak the multiple authorship of the corpus. Al-Fārābī (died 950), a younger contemporary of al-Rāzī, wrote a work justifying alchemical secrecy on the grounds that unrestricted knowledge of gold making would destroy economies—a common fear throughout most of alchemy's history.[31] In contrast, early European alchemy, such as that described in the *Summa perfectionis*, is remarkably free from overt secrecy, even if Geber mimicked the initiatic style of Jābir at the start of his book. Another sign of alchemy's initial move toward openness in Europe is that the subject began to be incorporated into the curriculum of the new medieval universities.[32] With the advent of controversy and criticism, increased public and official scrutiny, and legal sanctions, however, Latin alchemy retrenched and withdrew, becoming more secretive, more encoded, more allusive, and consequently more elusive.

This increased secrecy manifested itself partly in a renewed spate of pseudonymous works. Thus, even though the real St. Thomas Aquinas was ambivalent or skeptical about alchemy, in the fourteenth century (well after the Angelic Doctor's death) an allegorical work called *The Rising Dawn* (*Aurora consurgens*) was circulated under his name. New alchemical books were likewise written under the names of revered (and safely dead) figures like Albert the Great, Roger Bacon, Catalan philosopher Ramon Lull, and a host of others—including, amusingly enough, even ibn-Sīnā, whose denial of chrysopoeia had ignited so much controversy in the first place. (In fact, even the Persian's most anti-chrysopoetic sentence would wind up rephrased and attributed to pro-chrysopoetic authors as a "hint" about how to make the Philosophers' Stone!) Pseudonymity both legitimated these writings by attaching them to celebrated names and provided anonymity for their true authors.

Similar motives of legitimation lay, in part, behind the new connections between alchemy and Christianity that were forged at about the same time. The writings of John of Rupescissa and those attributed to Arnald of Villanova provide the best illustrations.

John of Rupescissa (or Jean de Roquetaillade) was born about 1310 in the Auvergne, in central France; he attended the University of Toulouse and then became a Franciscan friar.[33] In doing so he was influenced by the ideas of a branch of the order known as the Spirituals, who opposed the increasing institutionalization of the Franciscan order as it grew, claiming that it had abandoned the ideals and rule of its founder, St. Francis of Assisi (1181/2–1226). The Spirituals, who saw themselves as the true followers of St. Francis, embraced radical poverty and fiercely criticized church hierarchy and the more mainstream Conventual Franciscans. The Spirituals were also caught up in apocalyptic fervor and a fondness for prophecies, believing that the Antichrist was about to appear.

Ecclesiastical authorities viewed the Spiritual Franciscans with distrust and discomfort, and eventually suppressed them.[34] John himself was arrested in 1344 and spent the rest of his life in a series of prisons. While incarcerated he wrote most of his books (both alchemical and prophetic) and received inquiries from numerous visitors, including high-ranking clerics. Although John's works do describe various sufferings in prison, apparently his confinement was intended not to silence him (otherwise he would not have had access to parchment, ink, and books) but rather to keep a close eye on a potentially troublesome self-styled "prophet." John of Rupescissa's alchemical writings must have been extremely widely circulated and copied in their day, for they are among the most common manuscripts about the subject that survive from the fourteenth and fifteenth centuries.

It might seem incongruous that a man so fervently committed to the ideal of poverty would also devote himself to finding the secret of making gold. Yet at the start of his *Book of Light* (*Liber lucis*), written about 1350, John states clearly why he studied chrysopoeia and why he decided to write about it.

> I considered the coming times predicted by Christ in the Gospels, namely, of the tribulations in the time of the Antichrist, under which the Roman Church shall be tormented and have all her worldly riches despoiled by tyrants. . . . Thus for the sake of liberating the chosen people of God, to whom it is granted to know the ministry of God and the magisterium of truth, I wish to speak of the work of the great Philosophers' Stone without lofty speech. My intention is to be helpful to the good of the holy Roman Church and briefly to explain the whole truth about the Stone.[35]

True to his Spiritual Franciscan views, John says that the tribulation of the Antichrist is at hand, and that the church will need every form of help to withstand it; that help includes alchemy. John was not the only Franciscan who thought this way. The same concern about the coming of the Antichrist lay behind much of what Roger Bacon—also a Franciscan friar—wrote to the pope about sixty years earlier: the church will need mathematical, scientific, technological, medical, and other knowledge to resist and survive the assault of the Antichrist. We are well familiar with the use of science and technology for national security; in the case of John and Roger, we find a medieval precedent that includes alchemy as a means of ecclesiastical security.

John provides a detailed recipe for making the Philosophers' Stone. He holds that it is to be made from a specially purified mercury and a "Philosophical Sulphur." The idea that the stone, like the metals, is composed of Mercury and Sulfur would become a standard notion in European alchemy. The only problem lies in the intentional ambiguity of the names Mercury and Sulfur, both of which, acting as *Decknamen*, might conceivably refer to almost anything. But John is explicit that for him, the Mercury is common mercury carefully freed of its impurities, and the Sulfur is found within "Roman vitriol" (iron sulfate).

John first describes a series of sublimations of mercury with vitriol and saltpeter, followed by various digestions and distillations. Despite the apparently clear directions, however, his first step will not work in a modern laboratory if followed verbatim. The sublimate "white as snow" that John describes making is undoubtedly mercuric chloride; therefore, the starting mixture *must* have included common salt, but this substance is not mentioned in the list of ingredients. There are two possible explanations. First, John's saltpeter might have been quite impure and contained a large quantity of common salt. In fact, his *De confectione* contains an annotation toward the end that notes how crude saltpeter ordinarily contains salt, and gives a method for purifying it by fractional crystallization. The second possibility is that John intentionally left out the crucial ingredient as a way of preserving secrecy. If this is the case, then it is significant that the end of *De confectione* includes a rather out-of-place paragraph describing the general importance of table salt (*sal cibi*, or "salt of food"), its ubiquity, its use in purifying metals, and so forth, and then states that "the whole secret is in salt." Is this an example of the dispersion of knowledge?[36] Whichever of these two explanations is correct, the historical message is the same: alchemical recipes have to be read with care. Those that seem unworkable need not reflect negatively on the author's abilities or veracity, but might rather indicate a "hidden ingredient"—either something present as an unsuspected impurity or something artfully omitted from the recipe.[37]

Once the need to include salt at the beginning is realized, a modern chemist can follow John's procedure quite far, and in fact should be impressed by the level of technical skill and practical knowledge he must have possessed. For example, John uses the concept of "mass balance"—the weight of the products of a reaction must exactly equal the weight of the starting materials—to prove that the "invisible Philosophical Sulphur" he wants to extract from Roman vitriol has in fact combined with the mercury.

The sign that the spirit of the vitriol is incorporated with the mercury is this: if you put in one pound of mercury you will still get the same amount back [as a sublimate], despite the fact that in being sublimed the mercury leaves behind many earthy dregs. This result would be impossible unless the mercury, whiter than snow [as a sublimate], had carried up with itself the purest spirit of the aforesaid vitriol, which is the invisible Sulphur.[38]

In other words, since the mercury loses the weight of its "dregs," it should weigh less than a pound as a sublimate, but the fact that it still weighs a whole pound means that the lost weight has been compensated by picking up the "invisible Sulphur" that John is striving to obtain. Thus, John makes use of the quantitative test of comparative weights to monitor and follow a substance that is otherwise "invisible" because it is never isolable, only transferred from one substance to another. Such close observation and monitoring of the weights of materials indicate a degree of clearheadedness and care in the laboratory that is often not attributed to alchemists. Given that modern chemistry recognizes that the transmutation of base metals into gold is not possible by chemical means, it has often been too easy to dismiss virtually everything the alchemists did and wrote in their pursuit of that goal. Nonetheless, the more that historians of science inspect alchemical writings carefully and contextually, the more impressive many of these writings become from a scientific and an experimental standpoint.

At a certain point, however, the results John describes no longer correspond to what modern chemistry would predict. The same situation is often encountered when reading alchemical procedures. Sometimes it marks a boundary where the author moves silently from something he has actually performed to something he predicts *should* happen. In other cases, it means that a necessary ingredient or operation has been silently omitted, or that we are not recognizing and interpreting an allegory or *Deckname* properly. There is also the possibility that the author's ingredients had a different composition from our modern equivalents, and so gave results we could not predict. (Chapter 6 addresses this issue further by exploring and exposing the chemistry hidden within early modern alchemical recipes.)

At each stage of his procedure, John references another alchemical author, Arnald of Villanova. The real Arnald of Villanova was a Catalan physician who was born about 1240 and died in 1311. Like John of

Rupescissa and Roger Bacon, Arnald had ties to the Spiritual Franciscans (although he was not a friar himself), and around 1290 wrote a book about the advent of the Antichrist that brought him into conflict with the theology faculty at the University of Paris, who took a very dim view of such prophetic utterances as opposed to their own rational Scholastic theology. Although many alchemical writings are attributed to Arnald, it is very unlikely that he wrote any of them. Some do show attributes of Franciscan spirituality, and some are similar in their approach and use of scripture to Arnald's own theological and medical writings—thus, his name was a reasonable choice to attach to them.[39] These pseudo-Arnaldian writings appeared throughout the 1300s, but only one of them certainly predates John of Rupescissa's work, namely, the *Tractatus parabolicus* (*Metaphorical Treatise*), which John cites by name in his *Book of Light*. This book forges a special connection between alchemy and Christian theology.[40]

The pseudo-Arnald holds, like John, that the Philosophers' Stone is to be prepared starting with mercury. But rather than providing the clear recipes that John does, pseudo-Arnald devotes his book to a comparison of the alchemical treatment of mercury to the life of Christ: "Christ was the example of all things, and our elixir can be understood according to the conception, generation, nativity, and passion of Christ, and can be compared to Christ in regard to the sayings of the prophets."[41] For Arnald, quotations from Old Testament prophets bear witness not only to Jesus Christ as the Messiah but also to mercury as the correct starting material for the stone. Just as Christ bore his torment in four stages—scourging, crowning with thorns, crucifixion, and thirst on the cross—so too must the mercury undergo a fourfold "torment" to be prepared into the stone. Just as Christ was glorified after His tribulation, so mercury is "glorified" by being turned into the stone. As the sufferings of Christ and His glorious Resurrection bring salvation and healing to the fallen world, so too the final chemical transformation of mercury into the Philosophers' Stone brings "healing" to the base metals, converting them into gold. There is probably also a silent comparison here with the Spiritual Franciscan view that the coming tribulations of the Antichrist will prepare the way for the establishment of a new age of peace.

Arnald's comparisons between Christ and mercury perform two functions. They offer allegorical language akin to *Decknamen*, and they serve to elevate the alchemical art by linking it metaphorically with the central mysteries of Christianity.[42] The prophets speak not only of the Messiah but also of chrysopoeia. The alchemical art is sanctified by association through the similitudes it bears to the life of Christ. Another early fourteenth-century author, Petrus Bonus of Ferrara, claims that such similitudes can work just as well in the opposite direction: a knowledge of alchemy provides knowledge (and even observable proof) of Christian doctrines. In his 1330 book, the *New Pearl of Great Price* (*Margarita pretiosa novella*), Petrus asserts that a knowledge of alchemy allowed the "ancient [pagan] philosophers" to predict the virgin birth of Christ by means of analogy with the preparation of the Philosophers' Stone. "I believe firmly, that should any unbeliever truly know this divine art, it would of necessity make him a believer in the Trinity of God, and he would believe in Jesus Christ our Lord, the Son of God."[43] The very title of Petrus's book links alchemy to Christ's parable of the merchant in Matthew 13:45–46. These linkages enhance the status of alchemy by transforming it into a kind of *holy* knowledge.

On a larger scale, the formulation of such linkages tells us something crucial about premodern ways of thinking. Specifically, premoderns tended to conceive of and visualize the world in multivalent terms, where each individual thing was connected to many others by webs of analogy and metaphor. This view stands in contrast to the modern tendency to compartmentalize and isolate things and ideas into separate disciplines. This crucial feature holds a key to understanding European alchemy more deeply; it is one focus of chapter 7.

While the pseudo-Arnald's *Tractatus parabolicus* provides the earliest-known extended linkage of alchemy with Christian theology, the two would thereafter remain close in many (but not all) alchemical writings. Crucially, John of Rupescissa makes it perfectly clear that allegorical texts like the *Tractatus* were read and deciphered for *practical* information.

> Master Arnald says that it is necessary to raise up the Son of Man in the air by means of the cross, which in literal terms means that the material that was digested in the third operation, after being ground finely, is put at the bottom of a flask to be dissolved, and the purest and most spirituous of what is there is then turned upwards into the air, and is raised up in the cross of the head of the alembic, like Christ, as Master Arnald says, was raised up on the cross.[44]

Thus, the reference to Christ's elevation on the cross signifies a chemical process of volatilization in which the prepared mercury is "raised up" by means of heat from the bottom of the flask into the "head of the alembic" (the highest part of the heating vessel), where the purified material condenses as a crystalline sublimate. "The Son of Man must ascend from the earth into the air, and ascend upon the cross of the alembic like a crystal."[45]

Medieval Latin puns strengthen these theological connections. The vessel used for subjecting metals to high temperatures and corrosives is still known today as a *crucible*, from the original Latin *crucibulum*, translatable as "little place of torment" and originating from the same Latin root (*cruciare*) as *to crucify*. (Recall that centuries earlier, Zosimos had envisioned his own processes as "tortures" of the metals.) Given the usual repertory of chemical operations involving melting, corroding, grinding, vaporizing, hammering, and burning, it does not require a huge imaginative leap to envision them as "painful torments" of material substances. Accordingly, in explaining the pseudo-Arnald's use of a verse from the gospels (John 12:24), John of Rupescissa writes, "Understand by the 'grain of wheat that must die in the earth' the mercury that must die in the earth of saltpeter and Roman vitriol."[46] Here the verb *die* puns on an alternate name for "mercury"—*argentum vivum*, literally "living silver," so called because the silvery liquid seems in constant motion, as if it were alive. Mercury's "death" thus comes about when it is converted into an unmoving solid, which is exactly what happens when it is ground together with saltpeter and vitriol and "disappears" into the powdery mixture.[47]

Alchemy and Medicine

John of Rupescissa wrote another alchemical work while in prison: *On the Consideration of the Fifth Essence of All Things (De consideratione quintae essentiae omnium rerum)*. With it, he extended alchemy into a new area—medicine.[48] During the Antichrist's reign, Christians would need not only gold but also their full health. Thus, John recounts how he sought a substance that could prevent corruption and decay and thus preserve the body from illness and premature aging. He found such a substance in the distillate of wine—what he called "burning water" or "water of life," and what we call alcohol. The Latin alchemical term for this delightful liquid—*aqua vitae*—lives on in the names of several liquors: the Italian *acquavite*, the French *eau-de-vie*, and the Scandinavian *akvavit*.

John considers this "burning water" the "fifth essence" of the wine, its *quinta essentia* in Latin. (*Quintessence* is a word still used to express the finest, purest, and most concentrated essence of a thing.) John borrows the word from Aristotelian natural philosophy, where it represents a substance different from and greater than the four elements (fire, air, water, and earth), namely, the imperishable and eternal material from which everything beyond the moon, such as the stars and planets, is made. The implication is that this terrestrial quintessence of wine is similarly impervious to decay. While this might sound outlandish, John almost certainly based his belief on empirical evidence—he notes how meat left in the open air quickly begins to rot, but when immersed in alcohol it is preserved indefinitely. He may also have noticed that while wine quickly degrades into vinegar, distilled alcohol remains unchanged. It is this stability and preservative power that John wishes to turn to medicinal use.

John was not the first to distill alcohol from wine; distilled spirits had been recommended medicinally by (the real) Arnald of Villanova. Interestingly enough, John writes that he identified alcohol as his sought-after preservative agent seven years after his imprisonment—in other words in 1351, by which time he had been transferred to the prison at the Papal Court in Avignon, where, it so happens, the distillation of wine for medical purposes had been carried out since the 1320s.[49] It is therefore highly plausible that he discovered and witnessed its properties there for the first time.

Yet John goes much further in his employment of this "water of life" than did earlier authors and practitioners. He describes not only its preparation but also its use for making medicinal tinctures. Some of these he produced by simply soaking herbs in alcohol; here he was quite correct that alcohol often works much better than water in extracting the medicinally active compounds from plant matter. John also goes beyond the usual repertory of herbal remedies used in traditional pharmacology in that he recommends using metals and minerals. Gold had long been believed to have therapeutic properties, especially for strengthening the heart, and John describes its preparation for use in alcoholic medicines. (Our modern word *cordial* for a liqueur derives from gold-based remedies for the heart; *cordialis* is the Latin adjective for things related to the heart, or *cor*.) Mercury, antimony, and other metallic substances were then, as now, generally considered toxic, yet John proposes the production of powerfully medicinal quintessences from them as well.

John of Rupescissa made medicinal preparations a key part of alchemical practice; alchemy (and chemistry) would forever after be closely linked to medicine, for both good and ill.[50] His writings exemplify the two major goals of later European alchemy—transmuting metals and preparing medicines. John believed that these twin aims promised the health and wealth that oppressed Christians would need during the reign of the Antichrist. The allure of both rewards persisted long after concern about the Antichrist's appearance had dwindled away. Similarly, while the use of Christian doctrine as a source of allegory, metaphor, and legitimation began with fourteenth-century alchemy, this dimension also continued to develop in succeeding centuries.[51]

The Pseudo-Lull and the Lost Crusade

John's twin goals of transmutation and medicine became more tightly interwoven over the next generation, when his ideas about the quintessence of wine were widely distributed under someone else's name. Soon after *Consideration of the Fifth Essence* began to circulate, another author—whose identity remains unknown—co-opted large sections of the book, combined them with additional materials, and produced the *Book of the Secrets of Nature or of the Quintessence* (*Liber de secretis naturae seu de quinta essentia*). This new author had more interest in chrysopoeia than in medicine, so for him, extracting quintessences provided a step toward preparing the Philosophers' Stone. Whereas John sought the incorruptible quintessence as a preservative of human health, the new author saw such incorruptibility as the logical starting point for producing a substance that could confer incorruptibility to *metals*, that is, transform corrodible base metals into incorruptible gold. The book circulated under the name of Ramon Lull or Llull (1232–1315), a Catalan theologian and philosopher of the previous century who actually wrote negatively about alchemy. In subsequent years, the list of alchemical works bearing Lull's name grew dramatically. Although the real Ramon Lull wrote none of them, many of these compositions contained enough features resembling his authentic writings that the attribution seemed plausible—and remained largely unquestioned—for centuries.[52]

Pseudo-Lullian writings comprise one of the largest and most influential groups of medieval alchemical texts. The longest of them, the
Testamentum (*Testament*), also appeared first—in 1332, a generation
before the *Book of the Secrets of Nature*.[53] Significantly, the *Testamentum*
itself never claims Lull as its author; it could scarcely do so, since it mentions dates after Lull's death. Nevertheless, the author of the *Book of the
Secrets of Nature* co-opted the *Testamentum* as part of the Lullian corpus
he began producing. The fact that the *Testamentum* contains characteristically Lullian elements and was written by a Catalan scholar made it
easier to reassign the originally anonymous work to Ramon Lull.

The *Testamentum* defines *alchemy* as "a hidden part of natural philosophy" that teaches three main topics: how to transmute metals, how
to enhance human health, and how to improve and produce precious
stones. The last of these topics is unusual for alchemical texts of the
period, so the *Testamentum* gives a recipe for dissolving small pearls into
a paste and then molding the paste into larger artificial pearls.[54] It also
contains recipes for medicinal waters. Most of this lengthy book, however, deals with making the Philosophers' Stone, which can provide precious metals, good health, and better gems by itself. The author of the
Testamentum holds that the stone is a medicine of universal application.
It "cures" base metals, turning them into gold; it removes the imperfections of gems; it cures all diseases for both human beings and animals, and
even stimulates the growth of plants.[55] The enormous popularity of the
pseudo-Lullian corpus codified the notion that the Philosophers' Stone
was the "medicine of men and metals." (Although, following Bacon, the
pseudo-Lull, and others, the stone was thought able to maintain human
health and thus prolong life, it was *not* considered an "elixir of immortality," as some popular treatments of alchemy claim.)[56] Interestingly,
the *Testamentum* also states that the stone can render glass malleable—a
supreme feat of technology, fabled and rumored since Roman antiquity.[57]

Legends about Lull the alchemist's life and his alchemical exploits
began to emerge in the early fifteenth century. According to the full-
blown legend current in the seventeenth century, he was converted from
skepticism about alchemy by his fellow Catalan, Arnald of Villanova,
who also taught him the secrets of the Noble Art. Lull then traveled to
England. Some versions say he was invited by one Cremer, the abbot
of Westminster and himself a frustrated alchemist who, while searching
for a teacher, found Lull in Italy. Once in England, Lull demonstrated

his abilities to King Edward, and said he could make enough gold for the king to launch a new crusade to recover the Holy Land. The king agreed to Lull's proposal, and set him up with a laboratory in the Tower of London, where he transmuted twenty-two tons of lead and tin into pure gold, which was then minted into new coins called rose nobles. But Edward double-crossed Lull; instead of using the gold as promised to finance a crusade, he used it to invade France, and Lull either was imprisoned or, in other versions, left England disgusted and depressed.[58]

Like most alchemical stories, this one accreted around some nuggets of truth. The *Testamentum* does refer to Arnald, and its colophon records that its author wrote it in London near the tower, so at least one of the authors accounted a "pseudo-Lull" was actually in England. He would have been there during the time of Edward III (reigned 1327–77), a king known to have supported alchemists, who *did* issue a new gold coin in 1344 called the noble, and who invaded France shortly thereafter. Yet none of these events could connect with the real Ramon Lull, who died when Edward III was three years old. (Some writers therefore endeavored to identify the deceitful king as Edward I or II.) Furthermore, the true rose nobles (bearing images of a rose and a ship) appeared only in the middle of the next century.[59] Despite all the problems with the legend, other alchemists routinely used the story of Lull's ill-fated dealings with the English king as a cautionary tale for their fellow chrysopoeians: keep quiet about your knowledge, and avoid the deceitful halls of power.

Further New Developments: Florilegia and Images

During the fourteenth and fifteenth centuries, the output of new alchemical writings continued to increase and diversify. The earliest works of Latin alchemy, such as Geber's *Summa*, were predominantly Scholastic in style: organized, logical, and straightforward—like textbooks. This format continued to be used until the seventeenth century, even as other—and ultimately more popular—styles emerged. One new literary form was the *florilegium*. The word literally means a "gathering of flowers," and it refers to a text that picks the choicest excerpts from a wide variety of books and arranges them into a "book of books." Florilegia are anthologies or compendia of short, informative quotations taken from many authors. These excerpts might present explanations of alchemical theory, or a series of cryptic sentences requiring interpretation, or recipes for various products, including the Philosophers' Stone itself. The florilegium format was not unique to alchemy; authors used it to organize materials and authorities for a wide variety of subjects in the late Middle Ages. Today, florilegia might seem boring or redundant, but we can imagine that in their day, when books were expensive and scarce, they played an important role in summarizing and disseminating information from a broad range of sources.

The late Middle Ages also witnessed the emergence of another alchemical genre—emblematic illustrations. Extended allegorical descriptions of alchemical processes and theories emerged in the Greco-Egyptian period, notably in the "dreams" of Zosimos. But in the fourteenth century, this allegorizing tendency, now firmly established within alchemy, went further by manifesting itself not just in metaphorical *words* but in metaphorical *images*.[60] The sophistication of such images ranges from simple woodcuts to artistic and technical masterpieces of enormous complexity. No popular book on alchemy today fails to reproduce an array of such images. Yet their beauty and allure can prove to be a double-edged sword; many modern writers have wrested them out of context, as if they were somehow independent of both their creators and the texts they were intended to illustrate—independent of the time, place, and cultural conditions under which they were produced. As a result, they have often been interpreted according to the whims of the modern viewer rather than according to the intentions of their original authors and the practices of their original readers. Emblematic images can tell us a great deal about alchemy, but only when treated historically and in context.

Probably the earliest text to incorporate allegorical figures is the *Rosarium philosophorum* (*Rose-Garden of the Philosophers*). Actually, several books with this title were produced in the 1400s and 1500s; the earliest of them is attributed (falsely) to Arnald of Villanova.[61] All of them are florilegia (hence the title), but only one is adorned with images. Curiously, these images appeared first as part of an independent German poem titled *Sun and Moon* (*Sol und Luna*) that was later spliced into the Latin prose text of the *Rosarium*. The *Rosarium*'s text was written first, during the fourteenth century, and the poem was written somewhat later but still before 1400. Whether the two are the products of one author or (more likely) two remains unclear. What is clear is that the poem and its images were used to better organize the original florilegium—each verse and image summarizes the theme of a section of text—and they probably functioned as a memory aid. The combined assemblage of Latin text, German poem, and woodcut images was first published in 1550.[62]

The *Rosarium*, as its title page proclaims, deals with "the true method of preparing the Philosophers' Stone." It begins with quotations regarding general alchemical themes and theories, the composition of metals, and the production of the elixir from the combination of two substances, here called Sol and Luna. To describe the combination of these two principles, it quotes a section of the *Turba philosophorum*, which counsels the reader to "marry your son Gabritius, dearest to you among all your sons, with his sister Beya who is a shining, smooth, and tender girl."[63] Here the personification of the two ingredients draws on Arabic—the name Gabritius is undoubtedly derived from *kibrīt*, the Arabic word for "sulfur," and Beya from *bayāḍ*, meaning "whiteness" and "brightness," surely referring to mercury. Thus, as with John of Rupescissa, the *Rosarium* presents the theory that the elixir is made from the combination of Mercury and Sulfur. The difficulty, of course, remains in identifying what "Mercury" and "Sulfur" actually mean in this context. The writer of the German poem *Sun and Moon* (here the Sun/Moon pair is equivalent to Sulfur/Mercury and Gabritius/Beya) has the Moon tell the Sun that he has need of her "like the rooster does of the hen," and the illustrator graphically depicts the "conjunction" of Sun and Moon as shown in figure 3.1. "Where there were two," the Latin text continues, "they are made as if one in body." The following illustration (fig. 3.2) accordingly shows the one body with two heads that results from the fusion of Sun and Moon.[64] Subsequent illustrations depict the departure of the "soul" from this hybrid (fig. 3.3), the cleansing of the dead body, the return of the soul to produce the first stage of the Philosophers' Stone, and so forth.

The images of the *Rosarium* are simple and straightforward, as befits pictorial summaries of a preexisting text. Some later examples of alchemical emblems are far more complex, however, and often function as coded—that is, intentionally secretive—communications that require the reader to apply a full measure of interpretative skill in order to grasp their meaning (chapter 6 illustrates how to do this). Yet even the *Rosarium's* simple images can strike the reader as shocking or outlandish. Sexual intercourse and reproduction are common elements of alchemical imagery, both textual and graphic. But given that alchemy is fundamentally a *generative* and *productive* practice (that is, it makes stuff), comparisons

CONIVNCTIO SIVE
Coitus.

O Luna durch meyn vmbgeben/ vnd susse mynne/
Wirstu schön/ starck/ vnd gewaltig als ich byn.

O Sol/ du bist vber alle liecht zu erkennen/
So bedarffstu doch mein als der han der hennen.

CONCEPTIO SEV PVTRE
factio

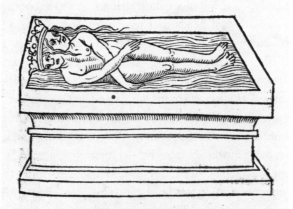

Hye ligen könig vnd köningin dot/
Die sele scheydt sich mit grosser not.

ANIMÆ EXTRACTIO VEL
imprægnatio.

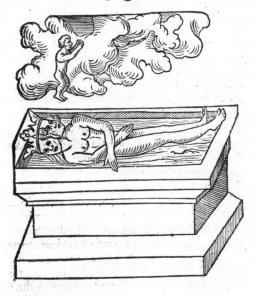

Hye teylen sich die vier element/
Aus dem leyb scheydt sich die sele behenbt.

Figures 3.1–3.3. Conjunction, Conception, and Extraction of
the Soul, emblematizations of stages in preparing the Philosophers' Stone.
The verses in German are from the poem *Sol und Luna.* From the *Rosarium
philosophorum* (Frankfurt, 1550). By courtesy of the Department of Special
Collections, Memorial Library, University of Wisconsin–Madison.

to procreation are actually appropriate. Alchemy's aim is to give rise to
new substances or new properties by combining existing ones, just as
parents give rise to new offspring through their union. Sex and sexuality
are among the most universal and common experiences of human beings, and so provide a ready source of similitudes and easily intelligible,
descriptive metaphors.[65] The idea or sight of two substances reacting and
combining to form a third easily suggests the image of a marital couple
to an imaginative mind practiced in drawing metaphors. Even modern
chemists frequently conceptualize reacting substances as pairs acting on
each other—no longer Mercury and Sulfur, but rather acid and base or
oxidant and reductant. Some of these modern pairs even continue to

suggest a kind of sexuality in their etymologies, such as electrophile and nucleophile, based on the Greek verb *philein*, "to love, kiss, or copulate." Even more unavoidable than taxes, death is also a common human experience, and was a part of daily life in the premodern world, though not in our sanitized and euphemistic modern society. Thus death, with all the attendant Christian doctrines of the departure of the soul and of final resurrection, appears in alchemical imagery as prominently as sex.

One exotic being commonly encountered in alchemical imagery but not in daily life is the hermaphrodite. Why would alchemists be seemingly so obsessed with beings that exhibit both male and female physiologies? In the *Rosarium*, a bicephalous hermaphrodite (fig. 3.2) results from the union of Sun and Moon. In some ways, this is quite sensible. Unlike with animals whose procreation produces offspring while leaving the parents intact, the combination of two material substances causes them to unite in a new, third substance with a new identity, losing their own independent identities in the process. Thus, the hermaphrodite actually represents something closer to alchemical processes. St. Albert the Great helpfully explains alchemists' use of this odd image with the clarity characteristic of the thirteenth century. In his book about minerals, the Universal Doctor explains the Mercury-Sulfur theory of the metals, saying that these constituents are

> like father and mother, as alchemical authors say when speaking meta-
> phorically. For Sulfur is like the father and Mercury like the mother,
> although it is more aptly to be expressed that in the commixture of metals
> the Sulfur is like the substance of the paternal seed, and the Mercury like
> the menstrual blood which is coagulated into the substance of embryos.[66]

The basis of this comparison lies in a well-established notion dating to ancient Greek medicine that males (just like Sulfur) are characterized by the qualities of hot and dry, while females (just like Mercury) are qualitatively cold and wet. In some substances, Albert continues, these pairs of qualities are not well segregated, and in such cases "it is to be observed that some hot-dry is joined to a wet-cold in the same complexion, and this complexion is hermaphroditic."[67] Thus, a hermaphrodite in alchemy represents a substance arising from the union of a substance that is "male" (hot-dry) and one that is "female" (cold-wet). Note also how Albert clearly distinguishes between the *metaphorical* usage of "father"

Figure 3.4. St. Albert the Great points to an alchemical hermaphrodite.
From Michael Maier, *Symbola aureae mensae duodecim nationum* (Frankfurt, 1617), p. 238.
By courtesy of the Department of Special Collections, Memorial Library,
University of Wisconsin–Madison.

and "mother" for Sulfur and Mercury, and a "more apt" comparison of
them to other *substances*—namely semen and menstrual blood—whose
literal combination (according to classical theories of generation) gives
rise to an embryo. Albert actually laments the fact that "proper terms"
do not exist for talking specifically about the production of material substances (particularly minerals), which is why, he explains, authors find
it necessary to discuss them using analogies.[68] His role in explaining the
meaning of hermaphrodites in alchemy was not forgotten by his successors. Indeed, two and a half centuries later he appeared in a seventeenth-
century alchemical book pointing in explanation to a hermaphrodite
(fig. 3.4).[69]

Only a small handful of alchemical works containing emblematic images appeared in the fourteenth and fifteenth centuries. These share some
of the procreative or sexual images of the *Rosarium*, but many are drawn
as well from theological topics. Accordingly, the printed 1550 version of
the *Rosarium* uses two images borrowed from the early fifteenth-century
Book of the Holy Trinity (Buch der Heiligen Dreifaltigkeit)—considered the
first alchemical text written in German—which depict the Coronation of

Figure 3.5. The Resurrection of Christ as an emblem of a step in
an alchemical process. From the *Rosarium philosophorum* (Frankfurt, 1550).
By courtesy of the Department of Special Collections, Memorial Library,
University of Wisconsin–Madison.

the Virgin and the Resurrection of Christ (fig. 3.5).[70] The verses under the resurrection scene read, "After my many sufferings and great torments, I am resurrected, clarified, and free from all stain," recalling the expressions of the pseudo-Arnald.

By the start of the sixteenth century, Latin alchemy had developed in many ways beyond the Arabic *al-kīmiyā'* Europe had acquired more than three centuries earlier. The Noble Art's ancient and central interest in chrysopoeia remained undiminished, and the search for the secrets of transmutation continued with increased vigor, aided by a wealth of new concepts, materials, and observations. In fact, multiple "schools" of chrysopoeia had developed by this time, each promoting particular starting materials or particular procedures, and basing themselves on various theories of metallic composition and explanations of how the Philosophers' Stone could bring about transmutation. Yet even though most texts falling within the category of *alchemia* dealt with metallic transmutation, that was by no means the sum total of the field. By 1500, alchemy also included the preparation of medicines, as practitioners promoted an expanding array of chemically produced or enhanced medicaments. Medical alchemy (also known as *iatrochemistry* or *chemiatria*) would expand enormously in the sixteenth century thanks to the influential writings (and rantings) of the iconoclastic Swiss physician Theophrastus von Hohenheim, commonly known as Paracelsus.

Humbler and less visible applications of *alchemia* also flourished. The recipe literature continued to develop as more workshops turned to chemical methods for producing a range of goods useful in arts and manufactures—salts, pigments, dyes, mineral acids, alloys, perfumes, distillates of various sorts, and so on. Alongside these industrious and productive activities, a wealth of new concepts about the hidden nature of matter and its transformations developed. Some stemmed from the quasi-particulate matter theory of Geber, others followed Aristotle more closely, and still others were totally new. The potential of human artifice and the secret workings of the cosmos remained fertile areas for study and for new ideas. At the same time, alchemy achieved an increasingly visible presence in early modern European culture, arousing both admiration and critique. Its ideas, metaphors, products, theories, practices, and practitioners attracted attention from artists, playwrights, preachers, poets, and philosophers. At the end of the fifteenth century, *alchemia* was entering its golden age. The sixteenth and seventeenth centuries, the age of Copernicus, Galileo, Descartes, Boyle, and Newton, the age often called the Scientific Revolution, would also prove to be alchemy's great age.

FOUR

REDEFINITIONS, REVIVALS, AND
REINTERPRETATIONS

Alchemy from the Eighteenth Century to the Present

If I were to follow a strictly chronological sequence, this chapter would delve into alchemy's greatest epoch: the sixteenth and seventeenth centuries. Instead, I am going to leapfrog over that golden age for the time being to recount transmutational alchemy's sharp decline in the early eighteenth century and its subsequent revivals, sometimes in strikingly new forms. Although it might seem confusing to violate chronological order in this way, there is good reason to do so. Most readers probably are aware of several common claims about alchemy—for example, that it is fundamentally distinct from chemistry, that it is inherently a spiritual endeavor or involves self-transformation, that it is akin to magic, or that its practice then or now is essentially deceptive. These ideas about alchemy emerged during the eighteenth century or after. While each of them might have limited validity within a narrow context, none of them is an accurate depiction of alchemy in general. Nevertheless, they have all been put forward as general "definitions" of the subject over its *entire* history. For much of the twentieth century, even many historians of science participated in this practice. Because these claims are so widespread today, and so distort and limit our view of alchemy, it is better to examine them now, before they can distract from our efforts to gain a more histori- cally accurate depiction of alchemy as it existed during its marvelous golden age.

The Disappearance of Chrysopoeia

Chrysopoeia flourished throughout the seventeenth century. Across Europe, a flood of books on the subject issued from printing presses. Distinguished scientific thinkers discussed and pursued transmuta- tion. Simple workshops and princely laboratories eagerly sought the secret to the process, while academic debates for and against its possi- bility continued undiminished. By the 1720s, however, transmutational alchemy found itself in sudden and astonishingly swift decline. By the 1740s, chrysopoeia was considered in most (but not all) places a relic of the past. Occasionally it retained historical or antiquarian interest, but it had largely become an exemplar of human folly. How did a thriving, 1,500-year-old endeavor come to be so suddenly discredited?

The precise reasons for so abrupt and rapid a decline are still being studied and debated by historians of science. A simple explanation that *seems* reasonable in hindsight is that metallic transmutation was shown to be physically impossible. But the historical record does not support this notion. No new systems, experiments, or evidence appeared in the early eighteenth century that could have allowed anyone at the time to conclude that chrysopoeia was impossible. What the record shows instead is an increase in often vicious attacks on transmutational alchemy specifically as something simply *fraudulent*. Such vilification was nothing new; it had been a companion of alchemy since both the Arabic and the Latin Middle Ages. In the early eighteenth century, however, something changed. The objections became louder, stronger, and more persistent, less focused on theoretical and rational arguments and more focused on the moral and social issues of fraud. The rhetorical mud slung at chryso- poeia and its practitioners suddenly began to stick like it had never done before.

Significantly, it was at this same time—the first decades of the eigh- teenth century—that the words *alchemy* and *chemistry* took on new and more restricted meanings. Previously, the two words had coexisted and remained largely interchangeable. Even when some distinction in their usage from that period is detectable, it is not consistent and only rarely the one automatically made today. For example, Andreas Libavius's famous 1597 book, titled simply *Alchemia*, describes how to perform chemical operations, use laboratory equipment, and make an array of chemical preparations—in short, what we would now unhesitatingly call chemistry—with little mention of gold making or the Philosophers' Stone.[1] On the other hand, the collection of treatises titled *Theatrum chemicum*, whose first edition appeared around the same time as Libavius's *Alchemia*, features dozens of chrysopoetic texts—exactly what we today would instantly call alchemy. The entire range of ideas and practices dealing with the production and manipulation of material substances and their properties—whether the making of gold and silver, or the making of medicines, dyes, pigments, acids, glass, salts, and so forth—could be, and was, called either alchemy or chemistry. The word *chemistry* was used more frequently, though, largely because of the recognition of *al-* as the Arabic definite article, and its subsequent removal as leftover baggage from *chemeia*'s passage through the Arabic-speaking world.[2]

Because these two words now carry a host of modern connotations (most often, that chemistry is modern and scientific, while alchemy is outdated and non-scientific), many historians of science have adopted the practice of using the archaic spelling *chymistry* to refer to the whole range of practices that nowadays would be classed under chemistry *and* alchemy. This terminology was suggested both to recognize the undifferentiated domain of "alchemy and chemistry" and to transcend the automatic implications prompted nowadays by the words *alchemy* and *chemistry*.[3] Take a moment to think about the associations instantly evoked when hearing each term. (Would you have bought this book if its title were *The Secrets of Chemistry*?) Now try to imagine getting roughly the *same* immediate impressions from both words. If you can do that—and it is not easy—you will begin to hear with the ears of most early modern listeners. In practice, however, it is easier to be jolted into remembering the shifting meanings of the two words by being confronted with the oddly spelled *chymistry*. Consequently, I will use the term *chymistry* hereafter where appropriate.

The redefinition of *alchemy* and *chemistry* arrived concurrently with the moralistic repudiation of metallic transmutation. I argue that the driving force behind these developments lay in large part in the desire to elevate the status of chymists and chymistry. Before the eighteenth century, chymistry suffered from a very poor public image, and chymists had an ill-defined, often unsavory, identity. Unlike physics, mathematics, and astronomy, chymistry had no established place in universities; it had failed to gain a foothold there in the Middle Ages. Nor was it ennobled by a classical pedigree, meaning that no respected authority of antiquity had written about it. Chymical work was often dirty, dangerous, and smelly (to say nothing of chymists themselves) and was something tied closely to artisanal labor. The figure of the chymist was called on as comic relief in seventeenth-century plays and literature, almost invariably in the role of a bumbler, fool, or fraud (see chapter 7). Chymistry's transmutational aspect carried the centuries-old association with counterfeiting, forgery, fraud, and avarice. Its medical and pharmaceutical aspects were usually tied to the practice of untrained "empirics," not to that of learned and licensed physicians. Even Robert Boyle (1627–1691), later to be proclaimed the Father of Chemistry for boldly championing the value of the discipline, felt obliged to apologize in the preface to his first book on the subject for devoting himself to "so vain, useless, if not deceitful a study."[4] While chemists today are rightly chagrined by the public association of their discipline with toxins, carcinogens, and pollution, their seventeenth-century predecessors suffered even worse problems of identity and status.

At the end of the seventeenth century, as chymistry continued to grow in importance and applications to scientific inquiry, medicine, commerce, and intellectual life, it began finally to be professionalized, developing the outlines of a formalized discipline. This professionalization occurred in many places, but is perhaps clearest at the Royal Academy of Sciences in Paris, a scientific society established in 1666. In 1699, five of the thirty places for members were earmarked specifically for *la chimie*, thus making the academy the first place where the subject achieved an official, high-profile, and state-supported status as an independent scientific discipline. As part of this newfound status, chymistry needed a makeover. It was necessary to clean up its rather sooty image so that it and its practitioners could achieve the prestige and respectability already enjoyed by the other sciences, and the bad public impression of the subject would not be extended to the Academy of Sciences. The secretary of the academy and chief crafter of its public image, Bernard le

Bovier de Fontenelle (1657–1757), already held chymistry in rather low regard, mostly because it did not have the "geometrical spirit"—that is, a neat system of deductive axioms, as in mathematics and physics—that he thought characterized "real" science. Chymistry's dubious reputation in the public mind just made matters worse. Government ministers overseeing (and providing funding for) the academy also made known their desire not to have chrysopoeia discussed within the institution. Thus, one part of chymistry's makeover involved quarantining transmutational activities, the source of so much ill repute, into a different category and severing all links with it.[5]

Accordingly, the Academy of Sciences issued some of the noisiest rhetoric condemning transmutational endeavors—not as theoretically impossible but simply as *fraudulent*. Everything within the ambit of chymistry that was most easily subject to criticism—for example, the Philosophers' Stone, metallic transmutation, and so forth—was split off and increasingly labeled as *alchemy*. The processes and ideas deemed useful (including, ironically enough, many theories that were developed in the context of finding the stone) remained as *chemistry*. Thus, much of what alchemists had actually been doing all along—probing the nature and structure of matter and studying and harnessing its transformations—remained as chemistry, even as the alchemists were condemned by ridicule. This strategy proved remarkably successful at the time, and remarkably invisible in hindsight. "Alchemy" became the scapegoat for chymistry's sins, driven from the respectable quarters where a newly purified chemistry could now reside. *Chemist* and *chemistry* became respectable terms—descriptive of modern, useful, productive, and "scientific" persons and things. *Alchemist* and *alchemy* became pejorative terms, descriptors of archaic, empty, fraudulent, even irrational persons and activities.

Scratching beneath the surface of the above sketch reveals a considerably messier picture than what first appears. Alchemy's public repudiation did not actually eradicate it but merely drove it underground. Many chemists—even within the Academy of Sciences itself—continued to work on the problem of transmutation. For example, the academy's chemist Étienne-François Geoffroy (1672–1731) published a paper in 1722, "Some Cheats concerning the Philosophers' Stone," in which he described the tricks and frauds used by pretenders to transmutation.

Geoffroy's paper was a key step in the academy's public repudiation of chrysopoeia, and has often been seen as marking the "end" of alchemy.[6] But Geoffroy actually cribbed most of his material from a book already a century old, written by a chrysopoeian alerting his fellow seekers after transmutation to sleights of hand they might encounter. Geoffroy's private library was crammed with books on transmutation, and some recently discovered manuscripts suggest that he continued to pursue transmutation experimentally (but quietly) well after his famous public denunciation of it.[7] Other chemists of the Academy continued their own chrysopoetic studies in relative secrecy well into the 1750s. There was no *scientific* reason for them not to have done so. But the climate generated by the moralistic attacks on transmutation and its separation from "legitimate" chemistry meant that respectable, professional chemists could no longer afford to be seen studying alchemy. Alchemy was therefore left for the first time without any high-profile public defenders.

The search for metallic transmutation and the Philosophers' Stone continued quietly, albeit on a reduced scale, and it continues quietly in some quarters to the present day. This continuing work was (and is) usually done privately, and so it is difficult for the historian to assess its real extent, save when a practitioner "goes public." The most famous of these latter instances in the eighteenth century involved James Price, a chemist and fellow of the Royal Society of London. In 1782, Price claimed success in transmuting mercury into silver using a white powder and into gold using a red one. He performed several transmutations before witnesses, and exciting news of his claims spread rapidly in the popular press both in England and abroad. The fellows of the Royal Society, however, were incensed at what they immediately denounced as "charlatanism." Here again, like at the French Academy of Sciences earlier in the century, the tight linkage of transmutation to fraud is evident, as is the keen sense of embarrassment felt by the society—some members wanted to expel Price immediately. Sir Joseph Banks, society president, demanded that Price demonstrate the process in front of other fellows to preserve the honor of the institution. Price initially demurred, saying that he had exhausted his supply of the powders and that it would take time and effort to produce more. Eventually, however, in July 1783 he invited fellows of the Royal Society to his home outside London for a demonstration. Sources disagree as to whether only three fellows bothered to show up or none at all, but what is clear is that on the appointed day, Price committed suicide by drinking poison.[8]

Alchemy and the Enlightenment

The repudiation of transmutational alchemy by the professionalizing discipline of chemistry was reinforced by broader trends during the period generally known as the Enlightenment (roughly the eighteenth century). Alchemy became one of the many foils writers of the era used to enhance the achievements of their own age and to distinguish it from everything that came before. Enlightenment rhetoric was full of stark polarities—of light replacing darkness, reason supplanting superstition, new thinking casting aside old habits. It dealt analogously with the new binary of chemistry and alchemy: modern, rational, useful chemistry supplanted archaic and misguided alchemy.

Accordingly, many eighteenth-century writers cast alchemy into a common bin with everything else considered unworthy of the so-called Age of Reason: witchcraft, necromancy, astrology, prophecy, magic, divination, and so forth—all classed under the catchall title of "occult sciences."[9] This conflation is clear in the title of a massive seven-volume collection put forth by Johann Christoph Adelung in the 1780s: *The History of Human Foolishness; or, Biographies of Renowned Black Magicians, Alchemists, Devil-Conjurers, Expounders of Signs and Figures, Fanatics, Fortunetellers, and other Philosophical Monstrosities.*[10] Undoubtedly, some alchemists of earlier centuries were involved in one or more of these other topics as well—but the vast majority were not. It is therefore erroneous to think of historical alchemy as ordinarily linked with them. Alchemy was neither magical nor a so-called black art. Most alchemists saw their practical work as operating completely in accord with natural processes, as recounted elsewhere in these pages.

Some promoters of Enlightenment ideals viewed the obliteration of chrysopoeia almost as a measure of their own success. Thus, Christoph Martin Wieland, editor of the monthly magazine the *German Mercury*, reacted melodramatically to reports of James Price's transmutations.

> I come now in sackcloth and ashes before the European public, and call forth all enlightened heads! Put on mourning clothes and pray to the

deities of true wisdom and of the Enlightenment that they may yet
smother in its birth this black misfortune that looms up against you.
Hear what I say! The ancient enemy of true wisdom, the huge old specter
of goldmaking, long assumed dead, rises up like Dedgial, the horrid
Antichrist of the last judgement, and endeavors to trample philosophy
and Enlightenment into the ground.[11]

Could metallic transmutation really have been such a threat? Wieland's
hysterical response illustrates how alchemy had become by the 1780s
an icon of everything "unenlightened." Just as early eighteenth-century
chemists came to define themselves through public opposition to "alchemy," those who defined themselves within Enlightenment rhetoric saw
a resurgence of alchemy as a threat to their own identity. These polarities
endured long after the eighteenth century; they form the background for
the shrill rejection by some late twentieth-century scientists and historians to the revelation that many iconic figures of science, such as Robert
Boyle and Isaac Newton, were deeply involved in alchemy.[12] The polar-
izing rhetoric of the eighteenth century made it seem impossible that scientific ability and reason could possibly coexist with alchemy.

Wieland asked the chemist Johann Christian Wiegleb (1732–1800)
to write a detailed exposition of every point in Price's accounts where
fraud was possible; Wiegleb's report filled twenty pages of the *German
Mercury*. By this time, he had already published his own *Historico-
Critical Investigation of Alchemy*, a review of the history of chrysopoeia
containing lengthy and vitriolic refutations of its claims. While criticiz-
ing alchemical ideas (both historical and scientific), Wiegleb also, like
Adelung, compared alchemy to witchcraft.[13]

But the Enlightenment was a complex phenomenon that spawned
diverse, even divergent, movements in different contexts. Thus, even
while chrysopoeia was being repudiated by some factions, it was being
adapted by others. Herein lies the reason that figures like Wieland and
Wiegleb continued to fulminate at such length against chrysopoeia: despite the attacks of the preceding half century, chrysopoeia was by no
means dead. In fact, the last decades of the eighteenth century witnessed
the first of several "alchemical revivals." In German lands, the number of
alchemical texts being published suddenly spiked in the 1770s and 1780s,
and several groups and journals (generally short lived) devoted to reviv-
ing, reorganizing, and pursuing chrysopoeia were founded.

PLATE 1. The coloring action of "water of sulfur" on silver. On the left, an untreated silver coin; on the right, an identical coin exposed to the water of sulfur described in the Leiden Papyrus. Author's laboratory.

PLATE 2. Explosions caused by heating sealed vessels and using unannealed glass were common in early modern laboratories. In the background, the chymist's wife silently comments on her husband's unsuccessful activities by wiping their child's bottom. Henrik Heerschop, *The Chymist's Experiment Takes Fire,* 1687; oil on canvas laid down on board. Courtesy of the Chemical Heritage Foundation Collections, Philadelphia; photo by Gregory Tobias.

PLATE 3. *Left*: a sample of stibnite, native antimony sulfide, the "antimony" of early modern writers. *Right*: the golden glass of antimony prepared by the author. *Top*: the "star regulus" of antimony, prepared by the author, showing the celebrated crystalline pattern on its surface.

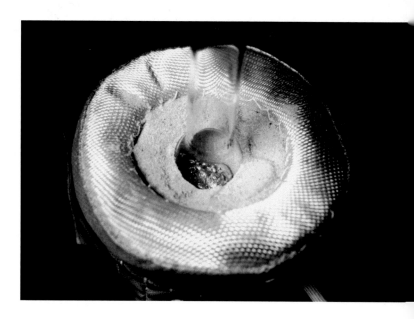

PLATE 5. The Philosophers' Tree grown in the philosophical egg. The short trunk and spread of branches are easily visible, and strongly correlate with Starkey's comparison to coral. Author's laboratory.

PLATE 6. Close-up of the Philosophers' Tree grown in the philosophical egg. The brilliant silver color and intricate ramification of the tree are obvious, as are the height and breadth to which the tree has grown; the amorphous starting material initially filled less than one-quarter of the belly of the flask (compare with plate 4). Author's laboratory.

PLATE 4. A mixture of Philosophical Mercury and gold sealed in a philosophical egg at the start of the process for making the Philosopher's Stone. Author's laboratory.

PLATE 7. A medal made from gold supposedly prepared from lead by transmutation in 1716. The reclining figure allegorizes the transmuted metal by carrying the attributes of Saturn (lead), namely the scythe and hourglass, while bearing the Sun (gold) as his head. The legend reads: "A golden offspring born of lead as parent." Courtesy of the Kunsthistorisches Museum, Vienna.

PLATE 9. Richard Brakenburgh, *An Alchemist's Workshop with Children Playing*, late seventeenth century; oil on canvas. Courtesy of the Chemical Heritage Foundation Collections, Philadelphia; photo by Will Brown.

PLATE 8. Adriaen van de Venne, *Rijcke-Armoede (Rich Poverty)*, 1632; oil on panel. Courtesy of the Chemical Heritage Foundation Collections, Philadelphia; photo by Gregory Tobias.

PLATE 10. After David Teniers the Younger, *The Alchemist,* seventeenth century; oil on panel. Courtesy of the Chemical Heritage Foundation Collections, Philadelphia; photo by Will Brown.

PLATE II. Thomas Wijck, *The Alchemist in His Studio,*
seventeenth century; oil on panel. Courtesy of
the Chemical Heritage Foundation Collections,
Philadelphia; photo by Gregory Tobias.

PLATE 12. Thomas Wijck, *The Alchemist,* seventeenth century; oil on panel. Courtesy of the Chemical Heritage Foundation Collections, Philadelphia; photo by Will Brown.

A key locus of this revival was itself a product of the Enlightenment—newly formed secret societies, especially in Germany, such as the Freemasons, Rosicrucians, and the constantly misrepresented and ephemeral Illuminati. Several such groups embraced alchemy in one way or another. Some Masonic lodges used (and continue to use) alchemical symbolism and language in their rituals. More dramatically, the German group known as the Gold- und Rosenkreutzer (Gold- and Rosicrucians), active in the 1770s and 1780s, established private and communal laboratories where their members pursued medical and transmutational alchemy experimentally. Many books in favor of transmutation—often new editions of sixteenth- and seventeenth-century classics—published in late eighteenth-century Germany were connected with Masonic or Rosicrucian organizations. Interestingly enough, these groups engaged with alchemy on a predominantly practical level—carrying out the same sorts of laboratory operations and experiments that characterized the chymistry of previous centuries.[14] Exactly how the linkage between secret societies and alchemy developed remains incompletely understood, but alchemy's long-standing tradition of holding ancient and privileged secrets harmonized extremely well these groups' claims of harboring ancient and arcane wisdom.[15]

The chemist Andreas Ruff provides an alternate late eighteenth-century evaluation of alchemy in which he voices his dissatisfaction with Wieland's and Wiegleb's brand of Enlightenment. In 1788, Ruff published a chemistry textbook dedicated to a Masonic lodge in Nuremberg. His volume is largely indistinguishable in content and style from other chemistry texts of the day, and would have proved as useful as any other to a chemical practitioner of the 1780s. At the end of his book, however, Ruff goes on to provide "Fundamental Rules" for the practical pursuit of transmutational alchemy, and offers a list of questions readers should use to assess the authenticity of anyone presenting himself as an alchemical adept. For Ruff, chrysopoetic alchemy remains very much a part of chemistry, as it had been throughout the prior century. He bemoans the weakened state of chrysopoeia, blaming it on the fact that

> we live nowadays in an "enlightened" world, in a time when every
> sixteen-year-old boy is already a champion of criticism and a persecutor of
> superstition and antiquity. They revile their forefathers who believed too
> much, who debated about many things they did not understand, and who

(to their shame) affirmed many things for which they could not declare a reason—save that they believed. Thus the grandfather is dishonored in his grave by his grandson, and the father by his son, and whosoever can say such things without being at all ashamed shall be proclaimed to have an "open mind."[16]

For Ruff, the scornfully dismissive attitude of the Age of Reason that ridicules whatever cannot be readily understood prevents people from investigating the unusual and the hidden—alchemy included. This prejudice threatens to deliver the world, not into enlightenment, but rather into a new "Egyptian darkness." Such uneasiness about the excesses of the Enlightenment forms a common trait among many late eighteenth-century supporters of alchemy. Other contemporaries launched their own critiques of the idolatry of reason, spawning the Romantic movement.[17] A certain rebellious or "anti-establishment" character often accompanied alchemy long afterward (it had already been an obvious feature of Paracelsian writings, with their vociferous attacks on the medical establishment). In the twentieth century, those skeptical or suspicious of "modernity" and its excesses sometimes turned to alchemy as an oppositional or countercultural stance.

Alchemy in the Nineteenth Century

The alchemical revival at the end of the eighteenth century lasted only a few years into the nineteenth. Nevertheless, once again alchemy was only dormant rather than deceased during the first half of the century. A scattering of publications about transmutation continued to appear, and their authors can be divided into two main groups. One group continued the traditions, methods, and approaches of seventeenth-century (and earlier) alchemy. Only a few publications from this group appeared in the early nineteenth century.[18] Toward the end of the century, however, Albert Poisson (1864–1893), a medical student in Paris, became enamored with traditional alchemy and convinced of its claims. He pursued it avidly in the laboratory and republished several alchemical classics, as well as his own book on the subject. Poisson planned a massive alchemical compendium that would have run to many volumes, but this project was cut short by his death at the age of twenty-eight from typhoid fever.[19] Publications by later practitioners likewise following the methods of early modern chrysopoeia continued to appear sporadically throughout the twentieth century, many of which continue to claim success in preparing the Philosophers' Stone or other alchemical arcana.[20]

A second group of nineteenth-century practitioners headed in new methodological directions. They continued to pursue metallic transmutation, but in new ways that often drew on contemporaneous scientific discoveries. In the mid-1850s, for example, the chemist and photographer Cyprien Théodore Tiffereau (1819–after 1898) presented a series of papers to the Academy of Sciences in Paris outlining how, while in Mexico, he had succeeded in turning silver into gold using common reagents. He maintained that the metals were actually compounds of hydrogen, nitrogen, and oxygen and were therefore interchangeable by altering the relative proportions of these components.[21] This idea is of course analogous to the ancient Mercury-Sulfur theory of metallic composition, but it also reflects chemical debates of the time. Recent discoveries had compelled many mid-nineteenth-century chemists to seriously reconsider the possible composite nature of the metals. Well-respected chemists who supported the compound nature of metals openly speculated that the alchemical dream of metallic transmutation might in fact soon be realized.[22] Thus, despite their estrangement in the eighteenth century, alchemy and chemistry did—in some periods—reestablish intellectual contact. One journalist expressed this striking mid-nineteenth-century rapprochement by writing in 1854 that "after having poured out so much scorn upon her, in our day chemistry is moving towards joining with alchemy."[23]

Under such conditions, the Academy of Sciences was more open to claims of metallic transmutation than it would have been previously. It not only invited Tiffereau to its assembly to present his results but also organized an official committee to examine his claims. Unfortunately for Tiffereau, neither he nor others could replicate his results in Paris. He returned to a quiet private life as a photographer. In 1889, however, he reemerged from obscurity, and began to give public lectures about his findings at which he displayed the gold he had produced in Mexico. The popular press ran excited columns about this "alchemist of the nineteenth century." In 1891, drawing on recent work in biology and microscopy, Tiffereau proposed that the transmutations he had observed in

Mexico were brought about by microbial action. He ascribed the failure of his processes in Paris to the absence the requisite airborne microorganisms that had been present in Mexico (near the precious metal deposits, where they ordinarily existed).[24]

On the other side of the Atlantic in the 1890s, an entrepreneurial chemist and mining engineer named Stephen Emmens offered the United States Treasury a method of turning silver into gold. Independent tests were made of his method (which involved hammering Mexican silver) both in the United States and in England, but the results were not encouraging.[25]

These examples of transmutational alchemy's continuation after its eighteenth-century "demise" probably form only the visible tip of the iceberg. Archival manuscripts bear witness to many more experimenters, and undoubtedly a much larger number left no trace of their activities. When writing his history of alchemy in 1854, Louis Figuier appended an entire chapter about hopeful mid-nineteenth-century practitioners. He noted the large number of them active in France, especially in Paris, described their ideas at length, and visited their laboratories.[26] There remain serious (and some not-so-serious) investigators at work on gold making today.

Tiffereau's memoirs and Figuier's book appeared—unbeknownst to them—at the dawn of a second revival of alchemy, one far broader and more influential than its late eighteenth-century predecessor. This second revival, which lasted through the entire second half of the nineteenth century and into the twentieth, was perhaps not so much a rebirth as it was a movement that promoted a radical reinterpretation of alchemy's entire history before the eighteenth century. It would significantly change the direction of alchemical thought and subsequent ideas about the subject.

Alchemy as Self-Transformation: Atwood, Hitchcock, and Victorian Occultism[27]

A new phase in the history of alchemy began in 1850 with the publication of *A Suggestive Inquiry into the Hermetic Mystery*. Its author was Mary Anne Atwood (1817–1910), a resident of Gosport on the English Channel, where she lived with her father, Thomas South. Atwood claimed that she and her father had uncovered the true meaning and practice of alchemy that had lain hidden in the secretive writings of former ages. Shortly after the book's publication, however, she purchased the entire print run and burned it on the lawn of her house. A manuscript work on the same subject in verse, titled *The Enigma of Alchemy* and written by Atwood's father, was cast into the same flames.[28] Only her personal copies of the *Suggestive Inquiry* and the few that had already been purchased or sent out to libraries by the publisher survived. Atwood's later followers claimed that this literary immolation was sparked by a "moral panic" resulting from a "realization of the sanctity of the Art" and a fear of being "betrayers of the sacred secret." But Atwood and her father might as well have saved their money, since the copies that did escape destruction were avidly read and circulated, and the work was reprinted several times in the twentieth century.[29]

The *Suggestive Inquiry* begins with a cursory history of alchemy from Egyptian antiquity to the seventeenth century, when, according to the author, "mistrust, gathering from disappointment" generated the "absolute odium" under which alchemy and its practitioners had lain ever since. Atwood asserts that the world is "fully ignorant of the genuine doctrine" of alchemy, because what *appear* to be laboratory operations are really nothing of the kind. The literal sense of alchemical writings is merely "wisdom's envelope, to guard her universal magistery from an incapable and dreaming world."[30] The balance of Atwood's treatise presents her thesis in ponderous Victorian prose, interwoven with a jumble of decontextualized quotations from alchemical and classical authors, and filled with obscure assertions, enraptured exclamations, and strangely distorted scientific notions. She claims to reveal the two main secrets of alchemy: the true starting material and the method for making the Philosophers' Stone. The starting material, she writes, is a ubiquitous, imponderable, intangible ether. The alchemical vessel for making the stone is the alchemist himself, who in a trancelike state can "magnetically" draw in this ether and condense it into the stone, which is a "pure Ethereality of Nature" or "Light inspissate," an incorporeal agent of universal change and exaltation that dwells within and enlightens alchemical adepts.[31] As Atwood declares, "Man is the true laboratory of the Hermetic Art; his life the subject, the grand distillatory, the thing distilling and the thing distilled, and Self-Knowledge [is] at the root of all Alchemical tradition."[32] In short, Atwood originated the notion that alchemy was a self-transformative psychic practice.

For Atwood, the alchemical process involves self-purification and offers the alchemist exaltation to a "higher plane of existence." The spiritualized adept not only controls the ether to produce the Philosophers' Stone within himself but can also manipulate ordinary matter by the same forces and thereby transmute lead into gold by a psychic, rather than a physical, operation. She declares that everything—be it mineral, vegetable, animal, or spiritual—can be exalted within its own sphere by the same power and process. She makes the bold claim that "no modern art or chemistry, notwithstanding all its surreptitious claims, has any thing in common to do with Alchemy."[33] Atwood's advocacy of alchemy and repudiation of chemistry as "merely physical" thus reinforced the division between alchemy and chemistry promoted more than a century earlier.

The source for Atwood's ideas lies not in late antique, medieval, or early modern alchemy but rather in her own time and place, specifically the English craze for Mesmerism during the 1840s. Franz Anton Mesmer (1734–1815), a Swiss physician active in Paris half a century earlier, had promoted a theory that an incorporeal fluid permeates the entire universe, connecting human beings to one another and to the rest of creation. The improper circulation of this fluid through the body causes illness, and certain individuals have the ability to control its flow using either their own bodies or magnets, thus acting as healers. The system came to be known as "animal magnetism," taking its name and some of its principles from contemporaneous studies of electricity and magnetism which were then conceptualized scientifically as the movement of "imponderable fluids."

Mesmer's system was widely investigated in France, but with ambiguous results. In 1784 Armand Marie Jacques Chastenet, marquis de Puységur, while practicing Mesmer's magnetizing principles on a young man, induced a trancelike state in which the subject exhibited a new personality and was supposedly able to read the thoughts of those around him. Puységur called this state "magnetic somnambulism," and it provoked huge debates that raged in scientific, medical, and popular circles in France for the next seventy years. In 1837, a French "magnetizer" arrived in England and began to perform public demonstrations. Thereafter and throughout the 1840s, animal magnetism attracted huge attention in Britain and spawned a flurry of debates, claims, and denunciations.[34]

It is only within this historical context that Atwood's *Suggestive Inquiry* can be properly understood. The "ether" at the foundation of her reading of alchemy is the incorporeal fluid of Mesmer's animal magnetism. The trancelike state necessary for the alchemist's self-purification and the concentration of the "matter" is the same as the "magnetic somnambulism" reportedly achieved by the practitioners of Mesmerism and, Atwood adds, also by the ancient Greek devotees of the Eleusian mysteries.[35] Analogously, in 1846 Atwood's father, Thomas South, possibly in collaboration with his daughter, published a small tract called *Early Magnetism, in Its Higher Relations to Humanity as Veiled in the Poets and Prophets*, which claims that the practice of Mesmerism is hidden allusively in the Homeric Hymns of ancient Greece. Besides bearing witness to the pair's enthusiasm for animal magnetism, this earlier tract also provides a precedent for their reading of Mesmeric subtexts into historical literature. It was after the publication of *Early Magnetism* that father and daughter began to read alchemical literature, becoming convinced that they found the same Mesmeric principles hidden there, and that the alchemists' use of animal magnetism opened the way to extraordinary esoteric knowledge and practices. In Atwood's words, "Mesmerism, as it is mechanically practised in the present day, is a first step indeed, and this only before the entrance of that glorious temple of Divine Wisdom which a more scientific Handicraft enabled the ancients experimentally to enter, and from its foundation build up, as it were, a crystalline edifice of Light and Truth."[36]

Of course, Atwood's claims about what alchemists were really doing have as much historical validity as the notion that the Homeric Hymns really describe animal magnetism. Her work provides a fascinating glimpse of ideas circulating in mid-nineteenth-century England, and develops alchemy (understood in a broad sense) in new directions. As a historical interpretation or description of pre-nineteenth-century alchemy, however, it is simply wrong. All the same, her concept of alchemical practice as a self-transformative process involving special psychic states and incorporeal agents launched a revival of interest in alchemy during the Victorian period, but—and this is a crucial point—largely within the context of a broader and widespread "occult revival" that swept Britain and the rest of Europe in the second half of the nineteenth century.

Modified versions of Atwood's formulations (distanced from her reliance on Mesmerism) remain at the foundation of many popular views of alchemy today. Her division of alchemy into "exoteric" (the language of chemical operations) and "esoteric" (the hidden spiritual transformations) was adopted even by many twentieth-century historians of science, who generally remained unaware of the origins of this division.

A similar division of alchemy into an exoteric, physical language serving to hide an esoteric, spiritual message was propounded independently, and just shortly after Atwood, by Ethan Allen Hitchcock (1798–1870), a general in the United States Army. His brief *Remarks upon Alchymists* published in 1855 attempted to show "that the philosopher's stone is a mere symbol signifying something which could not be expressed openly without incurring the danger of an *auto da fé*." After an unfavorable review, Hitchcock issued the longer and more detailed *Remarks upon Alchemy and the Alchemists*.[37] Unlike Atwood's extravagant notions, Hitchcock's reading of alchemical texts is straightforwardly moral and Christian. He argues that alchemy is simply an allegorical description of the moral life. Like Atwood, however, Hitchcock asserts that the alchemists did nothing akin to chemistry, and that "*Man* was the *subject* of Alchemy; and that the *object* of the Art was the perfection, or at least the improvement, of Man. The salvation of man—his transformation from evil to good, or his passage from a state of nature to a state of grace—was symbolized under the figure of the transmutation of metals."[38] For him, the alchemists' quest was entirely religious. Philosophical Mercury represented a clean conscience free from immorality, and once obtained it led to the Philosophers' Stone, which represented the consummate moral and holy life. Hitchcock believed that human improvement comes not by psychic exaltation to higher planes of existence but rather through the practice of true religion and morality. According to his thesis, the true nature of the alchemical quest was hidden in secrecy, owing to the "intolerance of the Middle Ages . . . known to every one," and he asserts that an "open expression of [the alchemists'] opinions would have brought them into conflict with the superstition of the time, and thus exposed them to the stake." Unfortunately, he does not explain exactly how an exhortation to morality and piety would have been viewed as heretical.[39]

Hitchcock interprets alchemical substances, theories, and operations allegorically, like a preacher unfolding a scriptural parable or a scholar expounding literary imagery. In point of fact, seventeenth-century preachers *did* sometimes draw on contemporaneous chymical language, processes, and theories as metaphors when sermonizing on morality and spirituality. Themes such as purification and distillation readily lend themselves as moral or spiritual symbols, and some chymical writers themselves pointed to such connections (this topic will be covered in chapter 7). But while Hitchcock's view of alchemy does overlap in this way with some historical precedents, his interpretation of alchemy as no more than moralizing allegory is incorrect. Nevertheless, Hitchcock's ideas were frequently cited in the nineteenth century, and his notion of alchemy as a primarily religious quest persists widely today.

The Victorian occult revival is too complex and remarkable a phenomenon to be treated fully here. Nevertheless, the new versions of alchemy promoted by Atwood and Hitchcock played a key role in the movement, alongside magic, spirit invocations, séances, and the many other esoteric practices then in vogue.[40] These esoteric formulations were summarized in 1893 with the publication of *The Science of Alchymy* by William Wynn Westcott (1848–1925), the "Supreme Magus of the Rosicrucian Society in England, and Master of the Quatuor Coronati Lodge."[41] Westcott linked Western alchemy with his own "Hermetic" interpretations of the Kabbalah, alongside ideas derived from a broad range of sources, including grimoiric magic, Neoplatonism, Buddhism, and Hindu yoga. The inclusion of these latter two was undoubtedly provoked by the garbled Eastern mysticism of "Madame" Helena Petrovna Blavatsky (1831–1891) enshrined in the Theosophical Society she founded in 1875, of which Westcott was a member. In 1888, Westcott helped to organize the Hermetic Order of the Golden Dawn, a secret society that flourished for fifteen years. The group's seminal influence on the poet W. B. Yeats has been well established, but its wider influence on late Victorian society is only beginning to be fully recognized.[42] Similar occult-oriented secret societies that included alchemy were founded in France and elsewhere in Europe.[43]

By the end of the nineteenth century, the incorporation of alchemy into the broad spectrum of esoterica and the mythic history of secret societies had become a matter of course. A popular work on the Rosicrucians presented alchemical adepts as mysterious, ageless, semi-immortal wanderers endowed with knowledge and being far above that of ordinary men, and the same was done in the Masonic context.[44] These nineteenth-century linkages to the (generically labeled) "occult" have remained strong in impressions of alchemy ever since.

The most prolific Victorian author on alchemy was Arthur Edward Waite (1857–1942).[45] He wrote over twenty books on occult topics ranging from Freemasonry and Rosicrucianism to devil worship and the tarot. In them, Waite criticizes both Atwood and Hitchcock for ignoring the alchemists' practical laboratory work with material substances, whereby (according to Waite) they successfully produced a physical Philosophers' Stone able to transmute metals.[46] He characterizes his own view as a "middle course": the alchemists worked on physical processes, but these were only the corporeal manifestations of "a theory of Universal Development" that applied equally to human beings and to metals. This view once again repeats the exoteric/esoteric division that characterizes many late nineteenth-century views of alchemy. Calling alchemy both "physical mysticism" and "psycho-chemistry," Waite defines it as "a grand and sublime scheme of absolute reconstruction . . . the divinisation, or deification in the narrower sense, of man the triune by an influx from above." Accordingly, he envisioned an "alchemical transformation of humanity" that would provide "the perfect youth to come" by calling forth "the unevolved possibilities of the body and mind of humanity."[47] Thus, alchemy for Waite represents the means for a "spiritual evolution" of mankind as a whole into a higher form of being.

After a flurry of Victorian-era publications, Waite published nothing more on the topic of alchemy for nearly thirty years until 1926, when he published his last book, *The Secret Tradition in Alchemy*. There he makes a surprising about-face and concludes that "between the age of Byzantine records and the age of Luther there is no vestige" of anything in the historical record showing alchemy to be other than a "record of experimental physics."[48] He gives the reader little indication of how, why, or when he changed his mind, or even *that* he changed it. His dramatic transformation is one more mystery in the history of alchemy.[49] Yet his earlier writings proved far more influential—perhaps the most influential of all the many publications of nineteenth-century alchemy within occultist circles.

The view of alchemy as a self-transformative, meditative, or psychic process operating on the alchemist himself originated in the nineteenth century but still has wide currency. Esotericists around the world have continued to develop these notions to the present day.[50] Features of the Victorian spiritual/occultist interpretation often remain the default understanding of alchemy as a whole. The notions of a self-transformative or psychic alchemy obviously have inherent interest and importance as part of the long history of alchemy. However, the *historical* claims they make concerning the activities of alchemists of earlier eras are not valid, and so must be set aside if we wish to have an accurate understanding of alchemy during its golden age and before.

This "spiritual" interpretation also spawned further new forms of alchemy. Perhaps most notably, it mingled with (and encouraged) the older and continuing tradition of practical laboratory chrysopoeia that harkened back to pre-eighteenth-century models. François Jollivet-Castelot (1874–1937) provides one excellent example. He continued the practical gold-making quest initiated by his immediate predecessor and colleague Albert Poisson, but hybridized it with esoteric themes drawn from the occult revival, partly through his associate Gérard Encausse (1865–1916), known as Papus, who founded several occultist organizations in France. In 1896, Jollivet-Castelot founded (jointly with Tiffereau and others) the Société Alchimique de France, which published a monthly journal under his editorship from 1897 until 1914 and again from 1920 until Jollivet-Castelot's death in 1937.[51]

His first book, *How to Become an Alchemist* (1897), combines an interest in practical metallic transmutation with such occultist topics as the tarot, and features a preface by Papus. It also expresses two concepts foundational for his view of alchemy. The first is the unity of matter, a restatement of ancient monism (accordingly, he used the *ouroboros* on the cover of many of his publications); the various chemical elements are actually all modifications of the same underlying stuff. The second is hylozoism, the idea that all matter is alive; it evolves and develops no less than animals and plants. Jollivet-Castelot carried out extensive transmutational laboratory experiments, using mercury, metals, arsenic, antimony, and even the newly discovered radium. He considered himself in the vanguard of a new chemistry, the instigator of a new Chemical Revolution that would overturn the "misguided" one led by Lavoisier that brought forth the illusion of chemical elements. This "hyperchemistry" of the future would be born from the union of modern chemistry with the ancient knowledge expressed in alchemy and occultism. Jollivet-Castelot's later works vigorously critique the scientific establishment and attempt to organize a "synthesis of the occult sciences."[52]

Jollivet-Castelot's hybridization of practical laboratory work with occultist ideas continued in the work of a significant number of twentieth-century figures, and his Société Alchimique was joined by analogous organizations in Italy, Germany, and England. The Alchemical Society, active in London from 1912 until 1915 and officially linked to the Société, included as its members a strikingly broad combination of chemists, historians, and esotericists (including all combinations thereof). Its briefly active journal published an intriguingly diverse set of articles.[53] Undeniably, the years after 1900 mark another rapprochement of sorts between alchemy and chemistry. Once again, new scientific discoveries sparked a more sympathetic look at traditional alchemical claims. In this case it was the sequential discoveries, starting in 1896, of radiation, radioactivity, and elemental decay. As soon as elemental transmutation—by both the spontaneous decay of radioactive elements and the bombardment of lighter elements with radiation—became an established fact, occultists as well as practically oriented seekers after chrysopoeia seized on these discoveries as a vindication of the entire alchemical tradition—a few going so far as to claim that alchemists must have discovered radioactivity centuries earlier. From the other side, even some sober chemists hailed the newfound element radium as "a modern Philosophers' Stone," because its radiations could transform one element into another.[54]

Hallucinations and Projections: The Psychologizing View

The spiritual/occultist view of alchemy also fed into another highly influential interpretation: the psychological formulations of Swiss psychoanalyst Carl Gustav Jung (1875–1961). Jung claimed that alchemy "does not deal at all, or for the most part at least, with chemical experiments, but probably with something like psychic processes expressed in pseudochemical language."[55] Jung stated that the contents of the alchemist's psyche became unconsciously "projected" onto the materials in his vessels: "during the practical work, hallucinatory or visionary perceptions take place, which can be nothing other than projections of unconscious contents." In other words, in the course of laboratory work, the alchemists fell into an altered state of consciousness in which their unconscious mind produced hallucinations indicative of mental contents, states, and activity, not unlike the images experienced in dreams. Thus, Jung claimed that alchemy was really a description of the unconscious, and that the alchemists' "experience had nothing to do with matter in itself."[56] Alchemy's "true root" was to be found not so much in philosophical ideas and outlooks but rather in "experiences of projection of the individual researcher."[57]

Jung did not completely deny a role for laboratory experimentation in alchemy, but he asserted that the real object of alchemy was the transformation of the psyche. Since the psyche could project its contents onto any sort of matter, the actual substances employed by the alchemist were not crucial. Therefore, very few of the processes aiming to produce the Philosophers' Stone (or much of anything else) contain, in Jung's opinion, any recognizable chemical meaning. Alchemy's allegorical language consequently arose not as a means of concealment but rather because these images were the way in which the unconscious projected itself onto matter. Accordingly, the many names given to the starting material for the stone arose from the fact that "projection derives from the individual, and is therefore different in each case."[58] Contrariwise, the unity of symbols, figures, and images that appear in alchemical writings led Jung to believe that they were expressions of a *collective* unconscious, a supposed universal hereditary legacy present in the psychology of all mankind, and in a sense the psychic analogue of inherited instincts. He thus claimed to explain both procedural similarities and widely divergent particulars within a single theory of the mind.

Jung's ideas bear similarities to those of nineteenth- and early twentieth-century occultists. Both share the perspective that alchemy is primarily not about material transformations but rather about psychic transformations within the alchemist. Both see alchemy primarily as a means of psychic development. Both claim that the alchemist's true work occurs in an altered state of consciousness. This similarity should be no surprise; Jung studied Victorian occultism early in his career. His doctoral dissertation, *On the Psychology and Pathology of So-Called Occult Phenomena*, was based on the séances of his cousin Helly Preiswerk in which he participated. A. E. Waite's works, among others, circulated in Jung's Zurich Psychological Club in the 1910s. Jung also drew heavily from earlier studies of alchemical symbolism by Freudian psychologist Herbert Silberer.[59]

Just as Atwood's *Suggestive Inquiry* inspired a throng of followers who developed and modified her basic idea, so too did Jung's formulations. Perhaps the most prominent offshoot was initiated by the comparative religionist Mircea Eliade (1907–1986) in the 1930s. Eliade, like Jung, was influenced by various occultist movements. Like both Atwood and Jung, he portrayed alchemy as something concerned primarily with self-transformation, and thought that alchemists experienced an initiatic experience leading to "certain states of consciousness inaccessible to the uninitiated." While the alchemist might busy his hands with chemicals and metals, his real quest concerned the soul. Eliade wrote that "the alchemist, while pursuing the 'perfecting' of the metal, its 'transformation' into gold, pursued in fact his own perfection."[60] Eliade also added the notion that alchemy was defined by a cosmic view of the world and everything in it as alive—a vitalistic or hylozoic view akin to ideas promoted by Jollivet-Castelot and his circle. Eliade's views once again radically separated alchemy from chemistry.[61]

Multiple modern alchemical strains come together in the influential work of Israel Regardie (1907–1985). During his youth, Regardie associated with groups and figures deriving from the Golden Dawn, and later studied psychotherapy, which he practiced professionally thereafter. His 1938 *The Philosopher's Stone* combines the Jungian interpretation of alchemy with concepts drawn directly from the spiritual/occultist interpretation, simultaneously incorporating such diverse elements as Eastern mysticism, Kabbalah, hypnotism, and animal magnetism.[62] His syncretic (some would call it undiscriminating) approach maintains that alchemical texts from the whole history of alchemy were simultaneously chemical, spiritual, and psychological, but that their main purpose was to unite the "several constituents of consciousness" and to develop "an integrated and free man who is illumined."[63] In the 1970s, Regardie attempted to add the material aspects of alchemy to his synthesis by carrying out practical laboratory alchemy himself, with the result that he permanently damaged his lungs due to poor ventilation of the fumes produced.[64]

Jung's, Eliade's, and Regardie's formulations, as well as those from later works dependent upon them, continue to be maintained and published today—not only in a wealth of popular literature but also in some writings by historians of science and other academics. Nonetheless, these claims about alchemy's true nature are simply not supported by the historical record, and therefore—as influential as they were during the twentieth century in a range of contexts—they are now rejected by historians of science as valid descriptions of alchemy. An array of scholars approaching alchemy from many different disciplinary perspectives have come to the same conclusion.[65]

Back to the Sixteenth and Seventeenth Centuries

All the redefinitions and reinterpretations of alchemy described in this chapter sprang from specific historical contexts and currents. Accordingly, they need to be studied as products of their own eras. Although mistaken in terms of their *historical* claims about pre-eighteenth-century alchemy and alchemists, they nevertheless form an important part of the long history of alchemy, and proved influential to subsequent generations of artists, writers, and many others.[66] To be sure, these redefinitions and reinterpretations have significantly influenced the reading and historical analysis of earlier materials. But a *third* alchemical revival currently in progress is now fundamentally revising our understanding of the subject. What used to seem familiar is no longer so.

The first alchemical revival—at the end of the eighteenth century—attempted to revive the practice and pursuit of chrysopoeia and chemiatria along the same lines pursued during alchemy's thriving golden age. It opposed the denigration that alchemy had suffered earlier in the century. The second alchemical revival—beginning in the mid-nineteenth century—presented radically new interpretations of what earlier alchemists had really been doing. It attributed positive, self-transformative, and even grand and cosmic designs to the earlier alchemists, and as such can also be regarded as a response to the earlier vilification of the subject as foolish, fraudulent, or mercenary. The third alchemical revival—beginning toward the end of the twentieth century—is quite a different phenomenon, for it is taking place among historians of science and other academics.[67] Its goal has been to use more careful and critical historical skills to achieve a more accurate understanding of what alchemists were actually doing and thinking (and why) during the many diverse stages of alchemy's long and dynamic development, from Greco-Egyptian antiquity to the present day.

In line with the goals of this currently ongoing alchemical revival, I now jump back in time to the sixteenth and seventeenth centuries, to take a fresh and unbiased look at the alchemists of that period in order to learn what they thought, what they were doing, and how they influenced the rest of their contemporaneous society and culture.

FIVE

THE GOLDEN AGE

Practicing Chymistry in the Early Modern Period

B y the end of the Middle Ages, alchemy was a mature subject thoroughly established in Europe. The years from 1500 to 1700, known
as the Scientific Revolution or the early modern period, witnessed its
continuing expansion and development.[1] Alchemy's central goals—
achieving metallic transmutation, producing better medicines, improving and utilizing natural substances, understanding material change—
developed in many directions during this time. Aided by the printing
press, introduced in the mid-fifteenth century by Johannes Gutenberg,
alchemical texts appeared in greater numbers and guises, many of them
intentionally secretive, using allegory, *Decknamen*, allegorical images,
and dispersion to guard their secrets. Debates continued about alchemy's goals and promises, while new links to theological and philosophical concepts were forged. Theories to explain results and guide practical
research proliferated, and the number of practitioners increased substantially. The result of such explosive growth has been twofold. First,
alchemical ideas and practices expanded their cultural reach into an
ever-widening array of thinkers and doers. Second, the burgeoning

diversity of alchemy makes writing a comprehensive account of its early modern history virtually impossible, or at best premature, for alchemy diversified into so many facets that many remain to be explored.[2]

The following chapters therefore cover only a representative slice of early modern alchemy, focusing mostly on its two central goals: metallic transmutation and pharmaceutical medicine. These two topics did not constitute the entirety of the subject at that time, but together they do represent a majority of it. The next two chapters illustrate how chrysopoeia and chemiatria were grounded in coherent theories and observations, and demonstrate that many alchemical workers were astonishingly good experimentalists. This chapter presents the basic principles, goals, and assumptions of early modern alchemists. If you were a seventeenth-century alchemist, what did you know, do, and try to achieve? The next chapter reveals their actual practices, using careful interpretation of allegorical images and laboratory replications to reveal what lies behind secretive and seemingly impossible claims, including several about the Philosophers' Stone itself.

The interplay between theory and practice in early modern alchemy must be stressed. It is sometimes assumed that alchemists worked more or less "empirically," that is to say, rather haphazardly and without theoretical principles or critical observation. The reinterpretations of alchemy covered in chapter 4 all reinforce this impression by distancing alchemy from chemistry—separating it from serious, investigative laboratory work with physical substances—and thus from the history of science more generally. This outlook is mistaken. Early modern alchemy was an endeavor of head and hand, a meeting of theory and practice. It involved exploring, understanding, and manipulating material transformations in the natural world. It is fully a part, indeed a crucial part, of the history of science.[3] Previous chapters have described the synonymity of the words *alchemy* and *chemistry* before the eighteenth century, and the chemistry hidden within several simple alchemical processes, thereby helping to pull these terms back together. As a reminder of this point, I will now increasingly use the words *chymistry* and *chymist*.

The Basics: Metals and Metallic Transmutation

Seventeenth-century chymists, like their medieval predecessors, recognized seven metals: gold, silver, copper, iron, tin, lead, and mercury.[4]

They called two of them—gold and silver—"noble" because of their resistance to corrosion, their beauty, and their rarity. They labeled the remaining five base or "ignoble" metals. These chymists conceptualized metals as *compounds*, whereas we now know them to be *elements*. This compound nature meant that one should be able to separate metals into their constituent ingredients, although by the seventeenth century there was considerable difference of opinion about what those constituents actually were. Many continued to hold that metals were composed of differing proportions and/or qualities of the two ingredients Mercury and Sulfur, a concept dating to the Islamic Middle Ages. Figure 5.1 illustrates the formation of metals underground from the rising subterranean fumes of Mercury and Sulfur. Others adopted a newer view attributed to Paracelsus (see below) that postulated *three* ingredients—Mercury, Sulfur, and Salt. Still others, adhering more closely to Aristotelian ideas, claimed that all metals (or all substances) were composed of a single, common "prime matter" (*materia prima*) that could be endowed with different "forms." Prime matter itself had no qualities of its own and gave only substance and quantity, while the form provided all the qualities (color, hardness, and so forth) of the specific material. For Aristotle, prime matter was a concept rather than something that could be put into a bottle—it did not literally exist on its own. Chymists, however, who were more focused on actual laboratory practice, tended to view prime matter more tangibly and in more material terms. Prime matter for them, if it could be isolated, offered a kind of material blank slate on which any form might be imposed, thereby producing any desired substance.

Several other, less widespread systems coexisted with these formulations. The key point to remember is that the concept of metals as *compound* bodies undergirded the possibility of metallic transmutation. Altering the proportions, qualities, or pattern of composition of the component ingredients should alter the metals, changing one into another.

Belief in metallic transmutation also rested on observational evidence: it appeared to be a naturally occurring process. In mines, metals are rarely found in a pure state; lead ores almost always contain some silver, and silver ores contain some gold. This well-known observation suggested that base metals were constantly being transformed naturally underground into more noble ones, as their compositions were slowly

EX SVLPHVRE ET ARGENTO VIVO,
vt natura, sic ars producit me-
talla.

Figure 5.1. St. Thomas Aquinas (incorrectly thought to have written alchemical texts) points to the formation of metals underground from the combination of the exhalations called Mercury and Sulfur. The motto reads: "Just as nature produces metals from Sulphur and Mercury, so too does art."
From Michael Maier, *Symbola aureae mensae duodecim nationum* (Frankfurt, 1617), p. 365. By courtesy of the Department of Special Collections, Memorial Library, University of Wisconsin–Madison.

altered by the action of subterranean heat and water. Over hundreds or thousands of years, percolating groundwater slowly washed away the interfering impurities found in the base metals, while the gentle heat of the earth gradually cooked and digested the baser metals into the better decocted, more stable, and perfectly united composition characteristic of nobler metals. Therefore, the chrysopoeian needed only to find a way to do aboveground and quickly what nature was always doing underground and slowly.

Already in antiquity, each of the seven metals had become linked to a particular planet. If we include the Sun and Moon as planets (as pre-Copernican astronomy did) there are seven planets, just as there are seven metals (see chart below). The correlations between each pair varied somewhat during the first centuries of alchemy, but had become fixed by the time of Latin alchemy.[5] The origins of some connections are obvious: the two noble metals, for instance, are linked to the two major luminaries—gold with the Sun and silver with the Moon—on the basis of brilliance, color, and relative worth. Other pairings are less obvious. Iron is linked to Mars, probably because iron (in the form of armor and weaponry) is naturally associated with the god of war. It is ironic that in modern times the observable color of the Red Planet has been found to be due in fact to iron compounds. Copper is paired with Venus, since the goddess's home and the richest ancient copper mines were both on Cyprus—the island that accordingly provides the Latin word for "copper," *cuprum*.

Early modern chymists did not necessarily consider these links between the metals and the planets to be other than symbolic or analogical, though a few did suggest that planetary influences played a role in the formation of the corresponding metal under the earth.[6] Partly based on the correlation between the seven planets and the seven metals, the greatest of the naked-eye astronomers, Tycho Brahe (1546–1601)—who maintained a chymical laboratory in his Danish castle-observatory named Uraniborg—referred to chymistry as "terrestrial astronomy" or "astronomy below."[7] This relationship between "astronomy above" and "astronomy below" echoes the *Emerald Tablet*—"as above, so below"—tersely expressing the interconnectedness of nature as seen by early modern eyes.

Gold	Sun	☉
Silver	Moon	☽
Copper	Venus	♀
Iron	Mars	♂
Tin	Jupiter	♃
Lead	Saturn	♄
Mercury	Mercury	☿

Each pair of planets and metals was given a shared symbol, and chymists regularly used the planetary names to refer to the metals. Copper was often simply called Venus, and lead Saturn. Most chymists—not just those interested in chrysopoeia—used this nomenclature into the eighteenth century. It is curious to note that the old planetary name Mercury persists to this day as the official English name of the element we really should be calling quicksilver. Why mercury should be the lone survivor of the planetary nomenclature of metals remains an open question, although the centrality of this anomalous liquid metal to so many chymical theories may be part of the answer.

Metallic Transmutation: Particularia *and Anomalous Metals*

Early modern chymists interested in transmutation—which means most of them, to one degree or another—had two pathways to pursue: the *particular* and the *universal.* The particular method focused on a potentially enormous number of transmuting agents of various potencies and abilities, called *particularia.* The name, meaning "particulars," comes from the fact that these substances could transmute only particular base metals into silver or gold. Thus, one particular would turn copper into silver, for example, but would have no activity toward the other metals. The implicit contrast is with the *universal* transmuting agent, the Philosophers' Stone, able to turn *any* base metal into gold or silver. Particulars had the advantage of being supposedly easier to prepare, but this benefit was offset by their specificity and low potency. Accordingly, some chrysopoeians tell their readers not to bother with *particularia*, because the amount of gold or silver produced does not recompense the labor and materials involved. Nevertheless, many other chymists discuss them extensively, and one well-organized but anonymous proponent even provides his readers with balance sheets of the cost of materials versus the potential gross and net profits to be obtained from the manufacture and use of various *particularia.*[8]

The famous English chymist Robert Boyle (1627–1691) compiled recipes for *particularia* into collections he called his "Hermetic Legacy." In the preface to one such collection, he wrote that "the greatest number of Particulars are not considerably Lucrative unles made in great quantitys, yet there are some that being skilfully wrought, even in small quantitys, may enable a poor and industrious Artist especially if he be a single man to get a Livelihood, tho not to grow rich." Thus, Boyle saw the low transmutational potency of particulars as a virtue, because "these meaner particulars requiring many hands, Materials & Instruments to carry them on with profit will set many poor people at work & thereby relieve great numbers enabling them, or at least assisting them, to get maintenance for themselves & their distrest familys."[9] In other words, metallic transmutation by means of particulars might spawn alchemical "cottage industries"—legions of the poor industriously applying themselves to chrysopoeia and thereby gaining a modest livelihood. Although Boyle's philanthropic dream of alchemical poor relief was never realized, it serves as a reminder that early modern alchemy was not only an intellectual endeavor for scholars who theorized and wrote books about the subject. It was pursued also in less sophisticated forms by various artisans, entrepreneurs, and many others hoping for profits.

There was no uniform method or theory behind particulars. Most were supposed to be melted together with silver or a base metal to bring about transmutation. Others were to be fused with gold to produce a greater quantity of gold, a process known as *augmentation* or *multiplication*. In some cases these procedures may have worked by producing an alloy that could pass for pure gold. Yet other particulars were corrosive solvents (called *gradators*) supposedly able to transmute part of a metal dissolved in them. Some other solvents, by dissolving only one component of a metal rather than the whole metal, produced "modified metals"—strange metallic substances unlike any natural metal. In this way, some chrysopoeians endeavored to extract a "tincture" from gold that contained all the metal's characteristic color, leaving behind an anomalous white metal. Chymists considered this tincture the separated Sulfur of gold, sometimes called the *anima auri*, or "soul" of gold. This tincture was then to be used to "tinge" a white metal into gold. Some considered this material the sought-after potable gold, a powerful liquid medicine—perhaps a universal panacea—supposedly prepared from gold.[10] The problem with medicinal preparations of gold is that they decompose back into gold very easily—the sign of true potable gold, however, was that it could not be turned back into gold, because the gold had been sufficiently "dissected" to preserve only its therapeutic part. The *anima auri* was just such an extracted component of gold.

Several writers claimed success with such a process, but other chymists objected, pointing out that this method of transmutation by "transplantation" of the Sulfur of gold was useless from a financial standpoint, in that it required the destruction of as much gold to obtain the tincture as the tincture itself would produce.[11] In a similar operation, the influential Flemish chymist and physician Joan Baptista Van Helmont (1579–1644) claimed that he had extracted a green oil from copper that left behind an anomalous "white copper."[12] This experiment was part of the reason Van Helmont concluded that there must be two Sulfurs, one "interior" and one "exterior," in the composition of metals. If his extraction had succeeded in removing *all* the Sulfur of copper, the residue should have been a liquid Mercury, but the fact that a solid white metal remained behind indicated to him that there must be a more "interior" Sulfur, harder to extract, that keeps the Mercury of copper in the form of a solid (but white) metal. The exterior Sulfur provides only the color of the metal.

The modern reader who knows that metals are elements might wonder what these chymists were actually doing. A satisfying answer is hard to find. It would be easy—in fact *too* easy—to dismiss the accounts of anomalous metals as intentional frauds or the products of overactive imaginations. Some of these reports may be on the order of thought experiments—that is, illustrations of what *should* happen given prevailing chymical theories. But some accounts seem more concrete, such as the case of *luna fixa* made by the chymist George Starkey (1628–1665) in the 1650s.[13] *Luna fixa* was supposed to be a white metal having the appearance of silver, but displaying all the other properties of gold—great density, high melting point, resistance to corrosion by nitric acid, and so forth. Several witnesses privately recorded that they watched Starkey produce this anomalous metal, and that goldsmiths found the metal to test like gold. Moreover, these goldsmiths bought the strange metal from Starkey at forty shillings an ounce, more than eight times the price for silver at the time. What did Starkey produce? It is virtually impossible to say. But a white metal heavy as gold and resistant to nitric acid recalls the properties of platinum and related metals. Given the surprising ability of industrious early modern chymists to isolate substances present in only low concentrations, is it possible that Starkey acquired a sample of ore or other metallic materials naturally containing small amounts of such a metal and succeeded in separating it?

How to Make the Philosophers' Stone: Getting Instructions

Most aspiring transmuters aimed beyond particulars toward the Philosophers' Stone. When rightly prepared, the stone was supposed to have such extraordinary potency that it could convert thousands, even hundreds of thousands, of times its weight of any base metal into gold. Two major hurdles stood in the way of those seeking the stone: identifying the correct material(s) with which to begin, and then finding the correct practical operations for turning such material(s) into the stone.

Pursuing answers to these two fundamental questions was the mainstay of any stone-seeking chymist. There were several ways to get started. One was to run across, or more often to be approached by, someone with a recipe to sell. Obviously, this method of gaining information provided ample room for deceitful dealings, so two words could be used to sum up this state of affairs: "Caveat emptor" (Let the buyer beware). A recipe that seemed too good to be true probably was. Nevertheless, some vendors probably believed in all honesty that they possessed a workable recipe, but did not have the means to undertake what was typically a lengthy and expensive process to verify its legitimacy.[14] Such recipes, along with the services of their owner, were frequently offered to would-be patrons, usually rulers or wealthy private individuals.

The trade in recipes for transmutation (and for virtually everything else in chymistry) was brisk throughout the early modern period. Across Europe, recipes passed from hand to hand and were exchanged in letters, by word of mouth, and in manuscript collections. Although modern scholarly attention tends to focus on the more theoretically rich expositions published in books by scholars, it is important to understand that in any census of early modern chymical manuscripts, collections of miscellaneous recipes and processes will usually be found to predominate. Most early modern writers of learned texts were themselves collectors and traders of various chymical receipts. They served as important means of exchanging chymical information, results, methodologies, and ideas.

Empirical experimentation provided another option in the search for the Philosophers' Stone. Such investigation—ideally enlightened by practical experience and extensive knowledge of the properties of various materials—could prove to be a virtually endless labor given the variety of materials and potential pathways to be examined. Hence, the ancient Hippocratic statement about the art of healing came to be applied as well to alchemy: "Ars longa, vita brevis" (The art is long, but life is short).[15] Nevertheless, the enormous number of possibilities could be greatly reduced with educated conjectures based on observations and on theoretical considerations. Accordingly, most early modern seekers after transmutation did *not* cook up witches' brews of anything and everything that came to hand, hoping thereby to stumble upon the stone. Serious investigators guided their work with the theories and knowledge of their day, similar to the way that experimental and industrial chemists of today guide their own researches.[16]

Another method involved searching in books. The relationship to texts marks an important difference between early modern chymists and their modern counterparts. Seekers after transmutation invested the texts of the "adepts" (those who claimed success in making the Philosophers' Stone) with greater authority and treated them with greater patience than a modern scientist would ever do with his colleagues' publications. Because chrysopoeians believed that these adepts had actually prepared the stone and that their cryptic writings contained hidden clues to the process they used, careful textual study formed an essential and substantial part of their search. Of course, no book ever provided a straightforward "recipe" for making the Philosophers' Stone. The various methods of secrecy employed meant that every text required close and patient interpretation, and that every step was potentially a misstep. As the pseudo-Arnald of Villanova wrote, "I shall speak such that I may deride the fools and teach the wise."[17] Nevertheless, books and manuscripts promised clues to finding the right process—if only they could be properly interpreted. Thus, it was with a combination of textual study and practical experimentation that aspiring adepts endeavored to prepare the stone.

Chrysopoetic authors commonly claimed that "all the adepts say one thing," meaning that regardless of apparent differences of expression, all authentic authors agree on how to make the Philosophers' Stone. This viewpoint encouraged the production of the florilegia and compendia which juxtaposed quotations from various authors. It likewise inspired Sir Isaac Newton (1642–1727), an avid seeker after the secrets of alchemy, to compile his enormous "Index chemicus" in an attempt to piece together the entire secret by collecting and grouping together similar terms and expressions found in over a hundred books.[18] Nevertheless, in reality there existed considerable disagreement over the correct material to take in hand and what to do with it. Chrysopoeians can, in fact, be classed into various "schools" based on their opinions about the right starting material.[19] Unfortunately, some modern accounts have taken chymical writers too literally, and exaggerated chrysopoeia's homogeneity and consistency over time and place. This misapprehension gives the impression of a monolithic, static—even fossilized—discipline mired in repetition from one generation to the next. Such a portrait is far from the historical truth, as we've already seen in the previous chapters. While a careful reading of original texts does indicate points of commonality, it also reveals a broad diversity of approaches, theories, and practices. It also shows that ideas and methods evolved in response to practical experience.

There was yet another resort for those frustrated with experimenting and reading. Some chrysopoeians hoped to contact a more immediate higher authority to reveal the secret they desired. In the most mundane cases, this meant finding an adept willing to teach them. According to the legend surrounding the pseudo-Lull, Abbot Cremer, frustrated by the fact that "the more I read, the more I erred," traveled through Europe in search of an adept, ultimately encountering Lull in Italy.[20] The legend surrounding Nicolas Flamel claims that he, unable to decipher the meaning of an emblematic alchemical book, went on a pilgrimage to Spain to seek a knowledgeable person who could explain its meaning to him.[21] Indeed, travels to gain knowledge became a stock feature of fictional alchemical autobiographies. Yet seeking the advice of an adept is not restricted to fiction. Robert Boyle, for example, made numerous inquiries to his visitors and through his correspondence about the secret of transmutation. In 1678, he eagerly awaited the fulfillment of a promise that he would be inducted into a secret international society of adepts who would share their knowledge with him. Unfortunately, his high hopes were dashed, either because the castle in which the adepts were meeting was blown up by a bomb or because the whole affair was an elaborate confidence game.[22]

Other aspiring chrysopoeians aimed yet higher, attempting to contact angelic spirits or praying for divine revelation. Elizabethan mathematician John Dee (1527–1608), during his famed conversations with angels carried out through the medium of Edward Kelley, did not fail to inquire about the Philosophers' Stone, and Boyle himself recounted stories of contacting spirits where the topic of how to make the stone featured prominently.[23] The latter even suggested that angels and the stone bore some special affinity for each other. Many, and perhaps most, published chymical authors recommended prayers of petition, alongside diligent practical work with texts and furnaces, as one technique for gaining knowledge. This recommendation would have been natural for anyone in early modern Europe undertaking a difficult or important endeavor. Heinrich Khunrath (1560–1605) advocated not only prayer but also steps to contact angels and to provoke revelatory dreams. (The relationship between alchemical knowledge and the divine will be covered in more detail in chapter 7.)

Such methods, however, had their potential dark side. For some observers, dabbling in supernatural means of gaining knowledge was illicit, for it risked delivering the seeker not into the safe harbor of angelic assistance but into the damning power of demonic entities. Jesuit polymath Athanasius Kircher (1601/2–1680), while highly positive about the power and utility of chymistry in general, worried about the pursuit of metallic transmutation. To his mind, the task was so difficult that after years of fruitless labors, the frustrated chymist would inevitably turn to any source of help, including demons.[24] His earlier colleague, Martin Del Rio (1551–1608), believed that while "human industry and zeal" could in fact uncover how to make the Philosophers' Stone, in some instances such knowledge might be acquired by means of a shortcut with "the devil as teacher."[25] In short, long-term frustration and unchecked desires opened the door for demonic entities to take advantage of an obsessed chymist. This concern was voiced as early as 1396, when the inquisitor Nicolas Eymerich suggested that alchemists "easily fall into attaching themselves to and consulting with evil spirits . . . just as astrologers are much disposed to invoking and consulting demons when they are unable to accomplish what they desire."[26]

How to Make the Philosophers' Stone: Choosing Starting Materials

Most early modern chrysopoeians argued that since the goal of transmutation is to alter metals, it is necessary to begin with metals or metallic minerals. Mercury, whose liquidity marks it out instantly as a curiosity, garnered great attention. Gold was high on the list as well, at least theoretically, but its cost and lack of reactivity made it less attractive practically. The semi-metal antimony, with its curious chemical properties, its odd status of being "almost" a metal—it is shiny and fusible like a metal, but as brittle as glass and evaporates in the fire—and its ability occasionally to produce a mysterious "star" pattern on its surface during production, provided a perennial favorite (see plate 3). Some texts claimed that salts offer the correct starting point; some chose vitriol (iron or copper sulfate) and some, following the lead of the early seventeenth-century Polish alchemist Michael Sendivogius, focused their attention on niter (also called saltpeter or, in modern terms, potassium nitrate) or some more generalized "nitrous substance."[27]

A few seekers wandered outside the mineral realm into the vegetable or animal kingdoms. The Jābirian corpus had much earlier advocated the use of organic substances, and some of the earliest Latin authors followed suit. Eggs, hair, blood, and other organic substances were named as possible starting points; Roger Bacon advocated this route in the thirteenth century. But by the fifteenth century, nonmineral approaches had been widely rejected, even ridiculed, by most chymical authors.[28] Still, the use of organic materials persisted in some quarters. It was probably while trying to make the Philosophers' Stone that Hennig Brand discovered the element phosphorus in the 1660s by strongly distilling residues from human urine. Among less theoretically sophisticated workers (of which there was never a shortage) there remained a focus on excrement—both urine and feces—which at least had the advantage of being starting materials easy and cheap enough to acquire in large quantities in the streets of early modern Europe. The use of excrement as a starting material stems from an ancient axiom that the material of the Philosophers' Stone "is of cheap price and found everywhere" and "is trodden underfoot."[29] Already in the fourteenth century, John of Rupescissa roundly criticized those who interpreted the axiom to mean excrement: "The matter of the Stone is a thing of low price, findable everywhere . . . and many crude people [*bestiales*], not understanding the meaning of the philosophers, have sought it literally in feces."[30]

The confusion resulting from *Decknamen* aggravated disagreements over the proper starting materials. When a respected author cites the metal lead (for instance) as the place to start, what does he really mean: *actual* lead, or some other substance merely *called* "lead"? Thus, the "schools" just mentioned could not maintain clear or permanent borders. For example, the interest in vitriol as a starting material stems in

Figure 5.2. An emblem of the Philosophers' Stone with the vitriol acrostic as motto.
From *Von den verborgenen philosophischen Geheimnussen* (Frankfurt, 1613).

part from sixteenth-century advice in the form of the motto *Visita inte-
riorem terrae rectificando invenies occultum lapidem* ("Visit the interior of
the earth and by rectifying you shall find the hidden stone"), an acrostic that spells out *VITRIOL* (see fig. 5.2).[31] But does vitriol always mean
vitriol? When Johann Rudolf Glauber (1604–1670) used the motto, he
thought it did. On the other hand, the motto is sometimes connected
(erroneously) with writings attributed to the mysterious Basil Valentine.
Some writings under this name do stipulate real vitriol for the first steps
in producing the Philosophers' Stone, but others use the word *vitriol* in
such as way that it must be a *Deckname* for "antimony ore."[32] With the
benefit of historical insight, we can attribute some of this discrepancy
to the multiple authorship within the Valentine corpus, but original
readers expended massive efforts in trying to understand and harmonize
contradictory utterances such as these. Getting sources to agree—not
just those by a single author but also those by multiple authors—was a
time-consuming and frustrating process, yet one crucial to the practice
of early modern alchemy.

It became a common exercise for later authors to reinterpret earlier ones to show that what they "really meant" supported the present interpreter's own ideas. Thus, an author promoting metals as the correct starting material might reinterpret Sendivogius's mention of niter to imply not the salt usually meant by that name but some "nitrous principle" found within a particular metal. A particularly difficult *Deckname* might provoke a series of interpretations that continually varied over time, depending on experimental results and an author's own commitments. The fifteenth-century chrysopoeian and Augustinian canon George Ripley, for example, wrote that "sericon" was the crucial starting material for the Stone—but what on earth was sericon? Some of the earliest texts imply that it was a lead oxide, probably litharge or red lead, while later readers—perhaps urged by the failure of lead oxides to produce the desired results—began to interpret it as a host of other substances.[33] Reinterpreting earlier writings sometimes required attributing secrecy and deception to authors who had no such intent. The Scholastic Geber, for instance, is remarkably clear and straightforward, but by assuming that he was more ambiguous and secretive than he really was, later authors could interpret him to justify a broad range of ideas that Geber certainly never had in mind. The result is that the "right meaning" of texts constantly slips out from under the feet of their interpreters. Stable places to stand while seeking the Philosophers' Stone can seem as hard to come by as a resting place for the boulder of Sisyphus.

The *number* of ingredients to use also provided a subject for debate. Many writers emphasized that only one thing was to be used—but the briefer and clearer the direction, the more open it was to interpretation. "One thing" might mean not "one substance" but rather "one class of substances" or "one mixture." Likewise, the basic substratum of all material objects—whether the prime matter of Aristotelians, the primordial chaos of Genesis 1:1, the water of Thales, or various other quasiphysical substances—could be considered "one thing." Thus, "one thing" could be cleverly (or deceptively) expanded into *any number* of discrete substances, since everything shares one single, ultimate, fundamental "stuff." Hence to say that the stone is made from one thing alone might be no more than a restatement of monism, which is of no help for practical work.

Most advice on making the stone, however, involved combining *two* substances (at least at the crucial stage). The early *Rosarium philosophorum* describes a binary mixture composed of two things variously labeled as King and Queen, Sun and Moon, or Gabritius and Beya. When the ingredients for the stone are not personified, the terms *Mercury* and *Sulfur* are most commonly used, and are sometimes specified as special (that is, neither the common substances by that name nor the components of the metals) by being termed *Philosophical* Mercury and Sulfur. The two-ingredient starting material sometimes goes under the name of *rebis*, from the Latin for "two-thing."

From a logical and practical standpoint, a binary mixture makes sense, since new substances generally arise from the interaction of two different materials rather than the transformation of just one. The combination of two substances also lends itself readily to analogy with biological sexual reproduction, where two parents are required.[34] In making the Philosophers' Stone, the hot-dry quality of Sulfur represents a "male" element, and the cold-wet of Mercury represents the "female." The binary analogies extend further, however. Common sulfur coagulates common mercury into solid cinnabar (mercuric sulfide, HgS), just as the dry Sulfur principle coagulates the wet Mercury principle into metals, or male semen coagulates female menstrual blood into a fetus. For chymical writers, Sulfur and Mercury represent a *pairing of complementary principles*: solid-liquid, dry-wet, coagulant-coagulated, form-matter, active-passive, and so on. Indeed, the terms *Mercury* and *Sulfur* must be seen as referring to two *groups* of substances (real or theoretical) identified by their *reactivity toward each other*. Similarly, modern chemists label mutually reactive substances with binary categories such as acid-base or oxidant-reductant. Strictly speaking, nothing is an acid in isolation, only in terms of its reactivity toward another substance that acts in the role of a base. One key difference of course is that modern chemists do not use this system to be intentionally obscure or misleading, although some introductory chemistry students might think otherwise.

Other numbers of ingredients also appear. George Ripley's widely read composition in rhyming verse, the *Compound of Alchymie*, seems to stipulate a mixture of *three* substances, and even gives their proper ratios to one another:

> One of the *Sonn*, two of the *Moone*,
> Tyll altogether lyke pap be done.

Then make the *Mercury* foure to the *Sonne*,
Two to the *Moone* as hyt should be,
And thus thy worke must be begon,
In fygure of the Trynyte.[35]

Ripley adds further to the confusion by implying that there are in fact three different mercuries required for the work![36]

How to Make the Philosophers' Stone: Women's Work and Children's Play

Compared with the overwhelming ambiguity surrounding the starting material(s), there was—at least by the early modern period—considerable agreement about the general method for turning such material(s) into the Philosophers' Stone. The prepared substance or mixture is placed in a glass vessel with an oval body and a long neck, often called the philosophical egg (*ovum philosophicum*) on account of both the size and the shape of its belly, and its function in "giving birth" (again recalling metaphors of reproduction) to the stone. The hopeful chrysopoeian then seals the neck of the flask to prevent the loss of volatile materials. This airtight closure of the vessel, often carried out by melting the sides of the neck together, was called the "seal of Hermes," in reference to alchemy's legendary founder. (The memory of this very practical step in making the Philosophers' Stone remains alive today in the expression "hermetically sealed.") The sealed "egg" is then placed in a furnace and heated (the correct temperature is another source of consternation). Heating sealed vessels is generally a bad idea, since there is no provision for the release of pressure as the enclosed air expands upon heating, and certainly there are many accounts of exploding apparatus (plate 2). This problem was even worse in the early modern period, since the glass vessels of the time were made with very thick walls and therefore more prone to cracking and thermal shock.

If the material has been rightly chosen and prepared, and a detonation avoided, in thirty to forty days the enclosed substance will turn black. Black, the first of the "primary colors" of the stone, was called the "head of the crow" (*caput corvi*), the "blackness blacker than black" (*nigredo nigrius nigro*), or simply *nigredo*, among many other names. This color marks not only the "death" and putrefaction of the substance but a happy sign that the procedure was correct. The *Rosarium philosophorum*, quoting Arnald of Villanova, exhorts, "When you see your matter turning black, rejoice: for that is the beginning of the work."[37] From this point onward, the continued application of heat carries the stone toward completion without the chymist doing anything but regulating the fire. Hence, this part of the process was sometimes called "women's work and children's play." But this duty in itself—like "women's work" in the early modern period—actually constitutes an enormous burden of labor, given the need to keep the heat constant for months at a time. Today, we do this effortlessly with the flick of a switch thanks to electricity and thermostats. The early modern chymist had only carefully sized pieces of charcoal added at regular and frequent intervals day and night, and the manipulation of the air vents on brick or iron furnaces, to maintain and control the heat. In an age before thermometers, the chymist had to rely on touch, sight, and smell to gauge temperatures.

Upon continued heating, the blackness is supposed to recede over the following weeks, replaced by a multitude of short-lived and often-changing colors, called the "peacock's tail" (*cauda pavonis*). Gradually, the semiliquid mass grows lighter and lighter, finally becoming a brilliant white, the second of the primary colors of the Philosophers' Stone. This whitening, albification, or *albedo*, marks the completion of the White Philosophers' Stone or White Elixir—a station on the way to the complete stone. At this point, the (now presumably very happy) chryso-poeian has the option of opening the vessel and removing some or all of the white material. After certain further treatments, including the addition of silver, this White Stone becomes capable of transmuting all base metals into silver.

Reaching the ultimate goal of the work, however, demands continued heating beyond the white stage. Most writers suggested increasing the heat gradually at this point, whereupon the white material turns yellow and then darkens to a deep red. The material has now, at last, advanced to its final stage and color, the Red Philosophers' Stone or Red Elixir. Once the long process of incubating the stone is complete, the flask can be broken open and the stone removed. As with the White Stone, a few more operations are necessary. The Red Stone must be "fermented" with gold—that is, mixed with real gold into order to transmute other metals into gold. It must also be "incerated"—that is, made as fusible as wax by the addition of a liquid principle, usually the Philosophical Mercury, so that it melts easily and can therefore penetrate metals and convert them successfully into gold. The completed Philosophers' Stone appears as a deep red, extremely dense, brittle, and fusible substance capable of penetrating the metals the way oil does paper.[38]

On the momentous day that the chrysopoeian is ready to carry out the longed-for transmutation, he takes a crucible, adds a base metal such as lead or tin, and heats it until the metal is in fusion (or heats mercury to near boiling). He then takes a small amount of the Red or White Stone, sometimes wrapping it in paper or a piece of wax, throws it into the crucible, and increases the fire. This process is known as *projection*, from the Latin word *projicere*, meaning "to throw upon." After a few minutes, after all the contents of the crucible are molten again, the product—gold or silver, depending on which stone was projected—can be poured out into an ingot. The amount of stone to use in this operation depends on its transmuting power. The freshly completed Philosophers' Stone is said to be capable of transmuting about ten times its weight of base metal, but the process of *multiplication* can greatly augment that proportion. By redissolving the stone in Philosophical Mercury and redigesting it through the three colors of black-white-red, a tenfold increase in potency is supposedly achieved. Reiterations of the process could then produce a transmuting agent of enormous power. According to one account, a sample of the Philosophers' Stone reputedly found hidden in a bishop's tomb by John Dee transmuted 272,330 times its weight of lead into gold.[39]

Explaining the Stone's Action

Chrysopoeians advanced various theories to explain the marvelous action of the Philosophers' Stone. All of them agreed that its action was purely natural, that is, operating by natural laws alone. It is important to stress this point, since moderns often imagine the transmutational process to be somehow "magical" or "supernatural." Although some *critics* tried to portray chrysopoeia as operating in a non-natural manner that involved demonic agency and trickery—and was therefore something to be avoided—virtually all its advocates insisted on purely natural explanations.[40] A range of common observations provided clear precedents for the transformative action of the stone. Every early modern person

knew that a little vinegar thrown into a barrel of wine would soon transform the entire quantity into vinegar. Likewise, a tiny amount of rennet
could coagulate many gallons of milk into cheese, just as a particle of
the Philosophers' Stone could coagulate a hundred or thousand times
its weight of mercury into gold. A small piece of leaven kneaded into a
far greater quantity of fresh dough would soon turn the entire mass into
leaven. Familiar natural events such as these provided concrete instances
of analogous material transformations no less surprising or powerful—although less lucrative—than the stone's action.

Some authors claimed that the stone functioned like an especially
powerful purging fire, burning out the impurities and superfluities
from base metals that prevented them from attaining the purity of gold.
Others held that it might be endowed with an excess of the Aristotelian
"form" of gold, and hence, when projected onto a base metal, could destroy the old form of the metal, replacing it with the form of gold. A
closely related notion was that the Philosophers' Stone carried qualities
like hot and dry (characteristic of gold) in extremely intense degrees,
and so a small amount of it could correct, for example, the coldness and
wetness of a large amount of lead. Another variation on this theme was
that the elixir is "plusquamperfect," that is, in regard to the mineral realm
the stone is *more than* perfect—it is gold elevated high above its usual
rank of perfection. Thus when mixed in due proportion with an imperfect metal, the imperfection of the metal and the plusquamperfection
of the stone average out to perfection, that is, to gold.[41]

Still other authors asserted that the stone contained a "seed" of gold
capable of transforming other metals into gold. Modern readers should
not interpret "seed" too literally and assume that this term implies an
organic or living substance. In the early modern period, the term *seed*
signified a powerful agent, an organizing principle, that works at the mi-
croscopic level to transform substances. Consider the origin of the metaphor in the vegetable realm. How does a plant convert water absorbed
from the earth into all the various substances found in plants, and then
organize those substances into the complex structures of leaves, flowers,
stems, and fruits? There must be some principle within the plant capable
of guiding these transformations to their proper ends, acting as both the
blueprint and the mechanism for carrying out the necessary transformations. Early modern thinkers, many of them well beyond the borders of
chymistry, called such organizing principles "seeds" (*semina* in Latin),

and considered them present not only in plants but also in animals and mineral substances.[42] For some, these "seminal" transmutations occurred through a reorganization of the elements or a rearrangement of the tiny particles of which the metals were composed.

Finally, some authors, and indeed no less a person than Robert Boyle, acknowledged that they could not give a thoroughly satisfactory explanation of the stone's action. But Boyle for his part noted that neither could anyone in his day give a satisfactory theoretical explanation of fermentation—yet he did not doubt the ability of brewers to make beer![43]

Chymical Medicine, Paracelsus, and Paracelsianism

The preparation of medicines had become a major part of alchemy by the early modern period. John of Rupescissa's mid-fourteenth-century notion of the preservative "quintessence" and his application of chymical methods to make better medicines from minerals, metals, and plants had been picked up and expanded by later authors such as the pseudo-Lull and thereby spread far and wide. The distillation of medicinal waters and essences, the production of medicinal salts, and the introduction of new substances and preparative techniques for pharmaceutical use all comprised an important facet of fifteenth-century alchemy. Much of this early interest in chymical medicine has, however, lain in the shadow of a looming sixteenth-century figure: Theophrastus Bombastus von Hohenheim, called Paracelsus (1493/94–1541), one of the most colorful characters of the early modern period.

Paracelsus spent much of his life wandering from town to town, generally stirring up trouble wherever he went with his iconoclastic and quick-tempered ways. It has been claimed, erroneously, that the word *bombastic*—in the sense of pompous speech—derives from his name. Paracelsus is best known as a vociferous critic of traditional medicine— his writings, frequently imitated in style by later followers, are filled with vitriolic and sarcastic condemnations of physicians, apothecaries, and the entire medical establishment. It is reported that he publicly burned the medical writings of ibn-Sīnā, standard texts for medical education at the time, as a sign of his contempt. Paracelsus's other provocative habits included lecturing (for the short time he gave medical lectures in Basel) and writing in his native Swiss German rather than Latin, and promoting the use of German medicinal plants over the more established classical Mediterranean ones. He was a strong advocate of *alchimia*, but only as one of the "pillars of medicine," that is to say, for its ability to prepare pharmaceuticals and to explain bodily functions. He showed no interest in chrysopoeia, and occasionally wrote contemptuously of it.

One of Paracelsus's innovations was his expansion of the two principles of metals (Mercury and Sulfur) by adding a third: Salt. Additionally, whereas the Arabic dyad applied only to metals and some minerals, Paracelsus extended his triad—called the *tria prima* ("the three first things")—as the essential constituents of everything. These three chymical "principles" provided a terrestrial, material trinity that reflected the celestial, immaterial Trinity as well as the human triune nature of body, soul, and spirit. Further, Paracelsus endeavored to generate an entire world system, embracing the whole of theology and natural philosophy, as an alternative to (and, he no doubt hoped, ultimately a substitute for) prevailing contemporaneous systems. For him, chymical processes provided the fundamental model for explaining natural processes in the physical universe as well as within the human body. For example, the cycle of rain through sea, air, and land was for Paracelsus a great cosmic distillation. The formation of minerals underground, the growth of plants, the generation of life-forms, as well as the bodily functions of digestion, nutrition, respiration, and excretion were for him inherently chymical processes. God Himself is the Master Chymist; his creation of an ordered world out of primordial chaos was akin to the chymist's extraction, purification, and elaboration of common materials into chymical products, and His final judgment of the world by fire like the chymist using fire to purge impurities from precious metals. Paracelsus's system has been called a "chymical worldview," and it proved remarkably influential in succeeding generations.[44]

Paracelsus believed that powerful medicines could be prepared even from poisonous substances using chymical means of separation, which he called *Scheidung* in his native German. Processes including distillation, sublimation, putrefaction, and solution could be used to divide a naturally occurring substance into its three primordial principles of Mercury, Sulfur, and Salt. He considered these three the useful and beneficial parts, and believed that their separation left behind the toxic "dregs" of the substance. Once purified, the *tria prima* could be recombined to yield an "exalted" form of the original substance, free from impurities and toxicity, and thus enabled to operate more powerfully and beneficially as a medicine. Always fond of inventing words, Paracelsus gave this process of separation and reintegration the name *spagyria*. The term has been explained as meaning "to separate and (re)combine," from the Greek words *span* and *ageirein*, meaning "to draw out" and "to bring together."

Among later Paracelsian commentators, the process of spagyria received explicit theological overtones. Upon death, the soul and spirit separate from the human body, which putrefies in the grave. At the end of time, when the world is destroyed by fire, a new spiritual body is raised up at the Final Resurrection, into which the soul and spirit, now purified from sin, are reinfused by God, thus creating a glorified, immortal, and perfected individual.[45] Analogously, the chymist uses fire to separate the volatile Sulfur and Mercury from the Salt, purifies each separately, and then recombines them into a wholesome and perfected substance. In this way, God Himself is a chymist at both the beginning and the end of time, and in reflection, the chymist's work to improve material substances is divinized, for he acts in a God-like capacity to improve the natural world. Some Paracelsians even held that all poison and toxicity entered the world only with original sin. Therefore, by using chymistry to purify now-poisonous substances into medicines, the chymist returned them to their wholesome, pristine, prelapsarian state as they were created by God in the beginning. In effect, the chymical process was thus *redemptive*, and the chymist participated as a co-redeemer of a fallen world.

Because Paracelsus was far from a clear or orderly writer—it was even claimed by one of his associates that he dictated all his treatises while drunk—and because few of his writings were published during his lifetime, his immediate impact was only moderate and localized. But in the second half of the sixteenth century, during the generation after Paracelsus's death, followers such as Adam von Bodenstein, Michael Toxites, Gerard Dorn, and Joseph Du Chesne (Quercetanus) collected and edited his manuscripts, thereby organizing, codifying, and generally reworking his often chaotic and contradictory claims.[46] It is through their work that Paracelsus became widely distributed across Europe. Because Paracelsus's legacy was so variously reconfigured by both admirers and critics, uncovering the real historical Paracelsus has proved extremely difficult. He became, perhaps more than anything else, a larger-than-life

figure of anti-establishment sentiment, an icon of intellectual and cultural defiance across scientific, medical, political, and theological spec-
tra.[47] This attitude harmonized well with those common to both the
Protestant Reformation and the Scientific Revolution, a feature that
partly explains why the figure of Paracelsus enjoyed greater popularity than the specifics of his ideas. Alliance with Paracelsianism was, to
be sure, distinctly more prevalent among those outside established
circles of orthodoxy—radical Protestants, medical "empirics" (meaning
those without formal training), and so forth. Consequently, multiple
"Paracelsianisms" emerged in the late sixteenth century and throughout
the seventeenth.

The various Paracelsians, as well as their critics, exerted a significant
impact on both medicine and chymistry. Yet it is important to realize
that not every advocate of chymically assisted medicine (chemiatria)
would have identified himself as a Paracelsian, nor did Paracelsianism
ever constitute the sum total of chymistry or even of chemiatria. In Italy,
for example, the distillation of medicated waters and spirits had a long
tradition before Paracelsus. Rupescissa, the pseudo-Lull, and others initiated and perpetuated independent traditions of chymical medicine
long before the Swiss iconoclast ever saw the light of day.[48]

Bitter controversies erupted between the various "Paracelsians" and
their detractors. Part of this criticism rested on disagreements over specific medical claims and practices. A key issue was the Paracelsians' belief
that proper chymical treatment could free poisonous substances of their
toxicity and turn them into health-giving medicines. Consequently, they
advocated the use of toxic substances like mercury, arsenic, and antimony as pharmaceuticals. For over a century, one boisterous treatise
after another, either praising or condemning the medicinal use of antimony, perpetuated the so-called antimony wars between Paracelsians
and their more traditional medical counterparts.[49] The battle in France
came to an end only after 1658, when Louis XIV, having fallen ill during
a military campaign and not responding to traditional treatments by the
royal physicians, was cured by a vomit induced with a dose of antimony
administered in wine by a local physician. The Parisian medical faculty,
the corporate body which had previously condemned the use of antimony, then had little recourse but to vote in favor of legalizing the use of
this Paracelsian *vin émetique*.

Another part of the criticism focused more broadly on the anti-establishment attitude of Paracelsians and their vision of chymistry. Many physicians quite reasonably objected to being publicly declared fools and having their formal training, scholarship, and licensing dismissed as worthless. But perhaps the most prolific (and invectively rich) opponent of all things Paracelsian was the pedagogue and chymist Andreas Libavius (1555–1616). Libavius took particular exception to the Paracelsian dismissal of formalized study and classical learning, and the attempt to circumvent educational, civic, professional, and social mechanisms. He decried the Paracelsians' neologisms and obscurantism; he mocked their writings as disorganized, inelegant, and imprecise. While sarcastically condemning a range of chymical authors, he simultaneously defended more traditional chymistry in all its aspects. He upheld chrysopoeia, its adepts, and its commitment to protective secrecy. He likewise promoted the practice of chemiatria, attacking physicians who went too far while condemning Paracelsians and so rejected entirely the utility and dignity of chymistry. Libavius's massive *Alchemia* (1597, expanded 1606) organized hundreds of chymical preparations and laboratory operations, many of them directed at the production of chymical medicines. Writing as a fervent advocate of chymistry, he sought to protect its domain from those he saw as unworthy and disruptive interlopers, and endeavored to define legitimate chymistry from illegitimate while carving out an academic place for it.[50]

Other Ambitious Chymical Projects

The application of chymistry to medicine and the use of chymical ideas to explain cosmological themes form two key facets of Paracelsus's work. Yet there are other chymical endeavors often connected with his name. The tract *On the Nature of Things* (by Paracelsus and/or one of his followers) tells how chymistry can produce even living beings in the laboratory. The pinnacle of this art is the production of a human-like creature called the *homunculus* (from the Latin for "little human being"). Drawing on the ideas that new life begins with putrefaction and that seed will always endeavor to produce something, the author claims that if human semen is sealed in a flask and placed in a gentle heat to putrefy, after forty days it will begin to move and produce a living being with a human form. After forty weeks of being fed with a chymical preparation of human blood, this life-form will develop into the homunculus. Although it looks like a human child, the homunculus is endowed with great knowledge and powers. Since it is the product of art, the homunculus knows all arts from its birth. It is likewise endowed with special powers and gifts not shared by ordinary human beings, because its purity is unpolluted by the admixture of female elements. (As a comparison, the same text claims that if one takes menstrual blood instead of semen, and treats it in the same manner, the result is not a homunculus but a basilisk—a hideous creature so noxious it kills by its glance alone.)[51]

The possibility of producing life in the laboratory did not appear problematic for medieval and early modern thinkers. The spontaneous origin of life from nonliving matter was considered a matter of course— a rotting bull carcass produced bees, and putrefying mud generated worms and insects. Theologians interpreted Genesis 1:24, where God commands "let the Earth bring forth living creatures," to mean that God endowed the elements from the beginning with the power to generate living creatures on their own, and that God-given ability continues to reside within matter.[52] What did provoke a host of moral and theological issues was the notion of artificially producing a *rational* life-form akin to a human being. A combination of wonder and outrage accompanied tales of the homunculus to the seventeenth century.[53]

While few (if any) chymists showed serious interest in replicating so bizarre a recipe as that of the homunculus, the topic of chymical palingenesis—returning dead materials to some sort of life—attracted more attention. The germs of this interest lie once again in the Paracelsian text *On the Nature of Things*. There the author writes that plants and animals can be "resuscitated" by chymical means. The chymist is instructed to burn wood to ashes, mix the ashes with the watery and oily distillates extracted from the same kind of wood, and leave the mixture in a warm place until it becomes a slimy material. After this material is allowed to putrefy, it is buried in fertile earth, where it will eventually sprout into a tree of the same sort that provided the wood, but "more powerful and noble" than before.[54] This process is clearly a spagyric preparation of wood. The wood is subjected to *Scheidung* by means of distillation, and the three ingredients are the *tria prima* of the wood. The tree's "resuscitation" into a glorified form bears a clear analogy to the perfected nature human beings will enjoy after the final resurrection. The same text claims that a similar process can be performed on small birds.

A more widespread version of this "chymical resuscitation" comes, however, from Paracelsian Joseph Duchesne. He describes having seen two very different ways in which the form of a plant expresses itself after its destruction by fire. He recounts how a friend burned a large quantity of nettles to ashes, extracted the salt from the ashes with water, and after filtering the solution (called a *lixivium*) placed it in a basin by an open window on a cold night. The *lixivium* froze solid overnight; in the morning both Duchesne and his friend (as well as many other witnesses) saw the image of nettles, complete with roots, stems, and leaves, within the mass of ice. Duchesne viewed this experiment as proof that the residual salts of an incinerated plant retain the form and vital principle of the plant, and that this vital principle could re-express the form under the right conditions. He saw it also as a physical proof of the resurrection of the body at the end of time, after the world is destroyed by fire.[55]

While Duchesne (and others) were often able to repeat the experiment of the frozen *lixivium*, he also tells of a more striking experiment that he was never able to perform on his own; nor could any of the many other chymists his account inspired. A Polish physician from Cracow once showed him a series of sealed and labeled flasks that contained the chymically prepared ashes of various plants. When one of these flasks was gently warmed over a candle, a ghostly image of the complete plant grew out of the powdery ashes. When the heat was removed, the image slowly collapsed back into the ashes.[56]

For the rest of the seventeenth century and into the eighteenth, a number of chymists tried their hands in various ways at palingenesis.[57] Jesuit polymath Athanasius Kircher collected and published several recipes for it, and claimed to have shown a successful experiment at his museum in Rome to many visitors, including Queen Christina of Sweden in 1657.[58] Sir Kenelm Digby (1603–1665), who had visited Kircher, described palingenetic experiments to a learned assembly at London's Gresham College in 1660, where he also successfully repeated Duchesne's experiment with the frozen *lixivium* of nettles. He likewise reported success in the palingenesis of crayfish by means of an essentially spagyric process.[59]

Van Helmont was one chymist who flatly denied the significance of Duchesne's icy nettles: "the good man did not know first of all that when

any ice begins to form it produces serrated points like the shape of net-
tle leaves."[60] Yet Van Helmont inspired the search for another chymical
arcanum deriving ultimately from Paracelsus—the alkahest. Paracelsus
used the word *alkahest* for a particular medicine for the liver, but Van
Helmont uses the same word for a liquid able to dissolve any substance—
a material called "circulated salt" (*sal circulatum*) by Paracelsus. The alkahest holds a key position in Van Helmont's system.

Van Helmont's comprehensive and influential worldview unifies chymical, medical, and theological ideas. He rejected the elemental status of
the Paracelsian *tria prima* in favor of monism, arguing (like the ancient
Thales) that water is the basic material substratum of all substances. He
based this theory on the prominence of water in Genesis 1 and on laboratory experiments. In his most famous experiment, he planted a 5-pound
willow sapling in 200 pounds of soil and watered it for five years. At the
end of this time, the tree had gained 164 pounds, while the weight of the
soil remained nearly unchanged. Thus, Van Helmont concluded, water
alone was turned into all the various substances of the tree. The various transformations of water, he argued, are managed by *semina* (seeds)
capable of organizing water into other substances. Most materials can
be turned back into primordial water through heat and cold, thus establishing a continuous cycle of creation and destruction. Fire destroys
substances by turning them into *Gas* (a word Van Helmont coined from
chaos), a noncondensable substance more subtle than any vapor. *Gas*
rises to the upper parts of the atmosphere, where, exposed to extreme
cold, it returns to elemental water that falls with the rain.[61] The alkahest
performs this return to water more quickly and usefully.

Any substance heated with the alkahest is first decomposed into its
proximate ingredients (the *tria prima*), and upon further heating is reduced to water. As such, the alkahest promised to be the ultimate means
of carrying out chymical analyses—a crucial means of gaining knowledge
for Van Helmont and his followers. "There is no more certain genus of
acquiring knowledge," he wrote, "than when one knows what is contained in a thing and how much of it there is."[62] By stopping the process
at the right point, and distilling off the alkahest, the "first essence" (*ens
primum*) of the dissolved substance would be left behind as a crystalline salt. This *ens* contained the concentrated medicinal powers of the
dissolved substance, free from noxious properties, much like a spagyric
preparation, but (supposedly) far easier to prepare.

Van Helmont claimed to have prepared the alkahest, but declined to give more than allusive hints about how to do it. The promise of the alkahest provoked intense interest; many chymists struggled to understand Van Helmont's hints and prepare it for themselves well into the eighteenth century.[63]

The search for these chymical arcana coexisted with the two main branches of early modern chymistry, chrysopoeia and chemiatria. These grander designs were also accompanied by a host of workaday deployments of chymical techniques—assaying, smelting, metal refining, glassmaking, and so on. Chymistry's importance and applications grew throughout the seventeenth century as commercial and manufacturing endeavors became increasingly vital to Europe's economy. Otto Tachenius, a German chymist living in Venice in the mid-seventeenth century, provides a glimpse of chymistry's breadth in his diatribe against the slanderers of chymistry. He asserts that without chymistry there would be neither bricks, lime, nor glass with which to build houses; no inks, paper, dyes, or pigments with which to print and color; no spirituous drinks like beer and wine; no adequate medicines, salts, or metals. "But why do I spend time in mentioning these things," he concludes, "when there is not an Old Woman in Italy, but will inveigh against the opposers of this Art? For without it, it is impossible for them to find out any thing to Colour and Dye their Hair."[64] Thus, seventeenth-century "chymistry"—the unified field of what we today call alchemy and chemistry—stretched across a wide domain: from seeking the Philosophers' Stone, metallic transmutation, the alkahest, and other alluring secrets, to explaining natural functions of the body and the cosmos, to illustrating theological truths, to refining metals, making medicines, and preparing cosmetics.

The first items in that catalog of early modern chymical activities provoke several questions that may be nagging some readers. Why did so many people believe that chymical arcana like the Philosophers' Stone really existed? How could chrysopoetic authors give so precise a description of the preparation, appearance, and properties of substances like the stone and the alkahest? Are books about the Philosophers' Stone merely flights of fancy or exercises in juggling words borrowed from one book to the next, or did these writings contain an experimental, practical basis? To what extent can the adepts' coded language be deciphered to reveal laboratory practices? In terms of chemiatria, was there any practical basis for believing that toxic materials could be made medicinal? All these questions cluster around the central issues of what chryso-poeians and chemiatrists actually *did* in their laboratories, how they did it, and what they actually saw and accomplished. It is the task of the next chapter to address these difficult questions.

UNVEILING THE SECRETS

E arly modern chymistry embraces many topics that are usually regarded today as separate disciplines—chemistry, medicine, theology,
philosophy, literature, and the arts. As a consequence, the question of
what its practitioners were really doing and thinking can be approached
in various ways. It is actually necessary to pursue several approaches in
parallel to capture the strikingly multivalent character of the subject.
One of these approaches involves chemistry. Even though early modern
chymistry was embedded in a network of assumptions, theories, aims,
social structure, and philosophical commitments that are very different
from those of modern chemistry, chemistry remains the modern discipline most closely akin to the chrysopoeia and chemiatria of the early
modern period. Both use operations of combination and separation to
transform matter and produce new materials.

   While it would be a serious mistake to reduce early modern chymistry (or alchemy, if you prefer) to some sort of a "protochemistry," common elements between what practitioners of the seventeenth century
did and what chemists of the twenty-first do nevertheless exist. What is

certainly true is that the ways substances behave and react have remained the same, even as the ways human observers explain and conceptualize them have changed. Hence, modern chemistry can help the historian of alchemy in two ways. First, a knowledge of the chemical and physical properties of substances can help the historian grasp the processes and ideas that early authors describe incompletely or allusively. Second, and more vividly, a working knowledge of chemistry enables the researcher to try to replicate—and thereby understand better and more deeply—historical processes and results. My experience, based on three decades of such work, is that historically informed laboratory work in replicating these processes can in fact provide otherwise unobtainable insight into the actual practices and content of alchemy and the activities of its practitioners. This chapter therefore calls in chemistry—specifically, the modern explanation and replication of early modern chemiatric and chrysopoetic processes—to help reveal sixteenth- and seventeenth-century practitioners' thoughts and activities.

Early modern chymists have not made it easy for historians (or anyone else) to figure out what they were doing. Many left no written records of their work, or these documents perished over time. Of the records we do have—thousands of books and manuscripts—many were written with such intentional obscurity and secrecy that they appear to offer little clear insight into the theories and practices of their authors. Even the early modern chymists often had difficulty understanding their colleagues' writings. After centuries of profound changes in culture and ways of thinking and speaking, it proves even more difficult to understand them. To be sure, for a long time the arcane style of chymical (especially, but not exclusively, chrysopoetic) texts discouraged serious scholarly inquiry altogether, or allowed ahistorical interpretations to flourish. Even seemingly clearly written sources often claim results that today seem impossible. Hence, many in the past have been led to conclude that such descriptions were not products of laboratory practice at all. Yet careful reading, aided by chemistry, gives a different picture.

Chemiatria: Impossible Results?

One of the most celebrated figures of early modern chymistry went under the unlikely name of Basil Valentine (fig. 6.1). Valentine presents himself as a native of the Upper Rhineland who became a Benedictine monk and

Figure 6.1. "Brother Basil Valentine, Monk of the Order of St. Benedict and Hermetic Philosopher," frontispiece to his *Chymische Schrifften* (Hamburg, 1717). On the table, the Philosophers' Stone (symbolized as a basilisk) rests in a philosophical egg. Courtesy of the Roy G. Neville Historical Chemical Library, Chemical Heritage Foundation, Philadelphia.

spent his spare time studying chymistry so that he could prepare medicines for his monastic brethren. As the popularity of his books increased,
more biographical details emerged, and by the end of the seventeenth
century an extensive life story had accreted around him. By that account
Valentine was a fifteenth-century monk at St. Peter's Abbey in the northern German town of Erfurt. His writings remained hidden for over a century. The longest of them, his *Last Will and Testament*, lay concealed in
the high altar of the abbey church, where its author hid it just before his
death. Some accounts claim that a stray bolt of lightning struck and shattered a pillar in the church, revealing the secret manuscript within—a
tale reminiscent of the account in the *Physika kai mystika*.[1] Yet another
story, told by the abbey's prior around 1700, claims that Valentine's manuscripts lay hidden in the wall of the abbey's refectory.[2]

Both historians and chymists endeavored to learn more about
Valentine—with inconclusive results. Modern scholarship suggests that
several authors are hidden behind the mask of Basil Valentine (a pseudonym most likely derived from *basileos valens*, a hybrid of Greek and
Latin meaning "powerful king"). The Valentine writings date no earlier
than the 1590s, although some may incorporate earlier material. One author is almost certainly Johann Thölde (circa 1565–1624), a salt manu-
facturer in central Germany who published the first five books printed
under Valentine's name.[3]

The most famous book in the Valentine corpus appeared in 1604 under the grand title of *The Triumphal Chariot of Antimony* (*Der Triumph-
Wagen Antimonii*). The first part is largely theoretical, while the second
contains about two dozen practical preparations, seemingly very clearly
described, based on antimony. Today, antimony is known as a fairly rare,
semimetallic element of moderate toxicity (sharing many properties
with arsenic), but for early modern chymists it was a source of inexhaust-
ible fascination.[4] Despite the toxicity of antimony compounds, most of
Valentine's preparations are pharmaceutical. (A later story claims that
the element's name derives from the effects of Valentine's preparations
on his Benedictine brethren: "anti-moine," that is, "against monks." This
etymology is amusing but spurious.)[5] The *Triumphal Chariot*'s emphasis
on transforming poisons into pharmaceuticals, and its vitriolic condemnations of the medical establishment, places it firmly in the tradition of
Paracelsianism.[6]

Figure 6.2. A schematic of Valentine's chymical transformation of poisonous
antimony into a medicine. At each step, toxic or inactive materials are supposedly
separated—an illustration of the Paracelsian principle of *Scheidung*.

The *Triumphal Chariot* applies the Paracelsian principle of *Scheidung*
to antimony to remove its harmful qualities and generate potent medicines. Valentine first describes a way to isolate the Sulfur of antimony
(fig. 6.2).[7] He begins by making "glass of antimony" (*vitrum antimonii*)—
a vitreous substance used commonly (and perilously) to induce vomit-
ing. He extracts the glass with vinegar to provide a red liquid, evaporates
the liquid to a gummy residue, then extracts the residue with spirit of
wine (alcohol) to give a sweet red oil. This oil is supposedly the Sulfur
of antimony, and is no longer emetic or purgative because all poisonous
qualities have been separated.

To the modern chemist, this account appears unlikely in the extreme. The toxicity of a poisonous element cannot simply be "removed."
Furthermore, no compounds of antimony are alcohol- and water-soluble
and red in color. Is this account, then, a mere fabrication, or perhaps an
imagined process based on Paracelsian notions but never confirmed in
practice? I decided that the best way to get solid answers to such questions was to attempt to make the "Sulfur of antimony" myself.[8]

Making the glass of antimony seemed a trivial preparation; this material appears commonly in early modern pharmacopoeia. Valentine even
apologizes for starting with something so easy, but the initial results of
replicating his procedure showed that the apology was unnecessary.
Valentine instructs the reader to grind antimony ore (stibnite, native
antimony sulfide [plate 3]), roast it slowly until it turns light gray, melt
this "ash" in a crucible, and then pour out the molten material to provide
"a beautiful, yellow, transparent glass."[9] Accordingly, I took antimony
sulfide and roasted it (laboriously, since it takes two or three hours of
gentle heating and constant stirring, a process known as *calcination*) to
a light gray "ash." This ash—predominantly antimony oxides—melted

only with great difficulty, and when poured out solidified to a dirty gray lump. Many repeated attempts, with modifications to the temperature and duration of roasting and the length of time the ash was kept molten, always gave the same miserable result. After having exhausted other ideas, I obtained a sample of ore from Eastern Europe (Valentine specifies the use of "Hungarian antimony"), ground it, roasted it, fused the ash, all *exactly* as before—and this time obtained the beautiful, yellow, transparent glass (plate 3).

What finally went right? Analysis of the ore showed that it contained a small amount of quartz, one of the most common minerals on earth. This minute quantity, about 1 to 2 percent of the ore's total weight, proved to be the key; without its presence, the glass does not form.[10] In fact, when I took the ugly gray lumps from the failed trials, remelted them, and added a pinch of powdered quartz (or silica, silicon dioxide), they also turned into beautiful golden glasses. The initial repeated failure of Valentine's recipe might reasonably lead us to conclude that his process was false or imaginary, or even that he was hiding a "secret." But when his *exact* conditions were reproduced—using ore rather than its pure, modern chemical "equivalent"—the process worked exactly as described. The impurity was crucial.[11]

Valentine next tells readers to powder the glass and extract it with vinegar to produce a red solution. Once again, the process failed. The yellow glass made with the addition of quartz gave no color to vinegar, even after weeks of stirring. The glass made from the ore gave only a pale reddish color after several days. Chemical analysis provided a surprise: this redness was due not to any antimony compound but rather to iron acetate, undoubtedly from trace amounts of iron in the ore. This red material was formed in such a small quantity that it seemed impossible for Valentine to have made as much of it as he claimed. This time the key lay in a disregarded detail in his recipe: Valentine writes that he stirred the roasting ore with an *iron* hook, and then stirred the molten glass with an *iron* rod. Antimony compounds corrode iron very quickly. Thus, Valentine's iron tools enriched his glass with iron compounds. They provided the very substance he was isolating as the "Sulfur of antimony." Valentine's Sulfur of antimony actually contains no antimony whatsoever; it was extracted not from the antimony but rather from his laboratory utensils!

This amusing conclusion explains Valentine's claims and observations perfectly. Vinegar dissolves the iron out of the glass, but also dissolves some antimony compounds—hence, the vinegar extract still has the purgative properties Valentine noted. But after the vinegar solution is evaporated, and the gummy residue is extracted with spirit of wine, only the iron acetate dissolves, leaving all the antimony behind in insoluble residues. As Valentine writes (correctly), "the residue that remains behind contains the poison, the extraction takes up only the medicine."[12] The alcohol extract is completely nontoxic, and is, just as Valentine says, "sweet," for iron acetate has a slightly saccharine taste.

Only a patient attempt to replicate Valentine's processes could have revealed these findings. They indicate that while his theoretical explanation of his process was flawed, he nevertheless recounted his observations with perfect accuracy. Although the results seemed implausible and the process was unworkable until the role of impurities was recognized, it is now clear that he described laboratory operations that he carried out and observed closely. To him, it certainly appeared that chymical treatment rendered a toxic material harmless, apparently confirming Paracelsian theory. My replication proves that even some apparently unlikely chymical claims rest on real laboratory operations. These operations not only display technical proficiency but give support to theoretical principles as well. This example does not, however, involve the most difficult part of chymistry to explain: metallic transmutation. Thus, it is now time to turn to that topic.

Chrysopoeia: Deciphering Hidden Knowledge

Can underlying chemical meaning and accomplishments be found even in allegorically written texts dealing with transmutation? Do the fabulous accounts and outlandish emblematic illustrations of bursting toads, copulating couples, and flying dragons have practical significance? Since the mid-nineteenth century, most interpretations of "alchemy" have either dismissed these allegorical texts or tried to explain them with ahistorical conjectures that almost never involve chemistry. It is amusing that the methods chrysopoeians used to veil their meaning from all but the most clever readers have continued to work so well, probably better than they ever imagined. Yet chrysopoetic texts were written not only to conceal but also to *reveal*. Chemistry can help us better understand these texts.

Basil Valentine's first book—*Of the Great Stone of the Ancients*—provides a good case study.[13] The first half presents general principles and cryptic advice about the Philosophers' Stone. The second bears the subsidiary title of *The Twelve Keys*, because it takes the form of twelve short allegorical chapters about the stone's preparation, "whereby the doors to the ancient Stone of our predecessors are opened."[14] Each "key" reveals (and conceals) one part of the process, meaning that if the reader could decipher the secret language correctly, he would presumably learn the whole procedure. Chrysopoetic texts often use a similar format of successive steps or stages to be deciphered. George Ripley, for example, wrote his fifteenth-century *Compound of Alchymie* in the form of twelve "gates," each one cryptically describing a single operation (such as solution, sublimation, putrefaction) necessary for making the Philosophers' Stone. Ripley himself adapted this format from the earlier Guido of Montanor, who described "steps" on a ladder to reach the stone, and Ripley's style was imitated by many later authors.[15]

Valentine's *Of the Great Stone* is an especially good example for study, since the 1602 edition added an allegorical woodcut to illustrate each key.[16] As with the *Rosarium philosophorum* (discussed in chapter 3), the text came first and the illustrations later. That is to say, in most (not all) original works that include allegorical images, the text is primary. The images therefore cannot be understood if they are taken out of their context. Unfortunately, it has been a common practice, especially in popular books and now on websites, to publish the images alone with little or none of the texts to which they belong. Thus, these images become loaded with various imaginative interpretations unfettered by such petty annoyances as historical context or authorial intent.

It will suffice to examine only the first three keys in detail. Figure 6.3 shows the image embedded in the first key. The corresponding text teaches that "all impure and contaminated things are unworthy for our work." The theme of purity continues with a comment about how physicians purge illness from sick bodies. The section relating directly to the image advises that

> the king's crown should be pure gold, and a chaste bride should be married to him. Take the ravenous grey wolf that on account of his name is

Der Erſte Schlüſſel.

Figure 6.3. The emblematic first key of Basil Valentine. From *Von dem grossen Stein der Uhralten* (Leipzig, 1602). By courtesy of the Department of Special Collections, Memorial Library, University of Wisconsin–Madison.

subjected to bellicose Mars, but by birth is a child of old Saturn, and that lives in the valleys and mountains of the world and is possessed of great hunger. Throw the king's body before him that he may have his nourishment from it. And when he has devoured the king, then make a great fire and throw the wolf into it so that he burns up entirely; thus will the king be redeemed. If this is done thrice, then the lion has conquered the wolf, and nothing more to eat will be found in him; thus is our body completed at the start of our work.[17]

The woodcut shows the king, his chaste bride, and the wolf (wearing a collar and looking rather more like a whippet) jumping over the fire. Paternal Saturn (identified by his crutch and scythe) stands nearby. What does it all *mean*? This riddle is relatively easy. The text clearly describes a purification process. In the context of metallic transmutation, the king is likely the "king of the metals," that is, gold. This gold (the king's body) is fed to a ravenous wolf who is a child of Saturn. In the standard planetary nomenclature, Saturn is lead; his child would then be something closely related, and useful for purifying gold. The answer is Valentine's favorite substance, antimony ore or stibnite. Stibnite was widely thought to be related to lead, and was used to purify gold.[18] Calling stibnite a ravenous wolf would make sense to anyone who has seen it react with metals. When melted, stibnite dissolves—"devours"—the metals with breathtaking speed. Corroboration comes from the hint "on account of his name [he] is subjected to bellicose Mars." In German, the name for the mineral stibnite is *Spiessglanz*, literally "spear-shine," in reference to its shiny needlelike crystals. A spear, like all weapons, is subject to Mars, the god of war.

This process works very well today. When a piece of impure gold (for example, a 14-karat gold ring or necklace, which contains 58-percent gold and 42-percent copper) is thrown into melted stibnite, it dissolves almost instantly. Metals other than gold are turned into sulfides that float to the surface. A brilliant white alloy of antimony and gold sinks to the bottom of the melt, where it is easily retrieved after the crucible has cooled. When this alloy (that is, the wolf with the king in his stomach) is roasted ("make a great fire and throw the wolf into it"), the antimony evaporates, leaving the purified gold behind. Now that the gold is pure, "nothing more to eat will be found in him"; thus, the "lion [king of beasts = king of metals] has conquered the wolf."

The second key discusses the beverages available in the "courts of the powerful," and notes how the "bridegroom Apollo," before his marriage to the "bride Diana," must be carefully bathed with water "which you must learn to prepare by various manners of distillation." Apollo is god of the Sun, and the Sun is linked to gold, so it is probable that this key starts with the gold purified in the first key. Gold, called king previously, is now called Apollo. *Decknamen* are not constant, even within a single book—the tricky (and perhaps playful) chrysopoetic writers multiply them unceasingly, sometimes within a single sentence. The author continues,

> The precious water in which the bridegroom needs to have his bath must be made most cleverly and carefully from two fighters (understand two contrary things). . . . It is not useful for the eagle to make his nest alone at the top of the Alps, for his young would freeze from the snow high up in the mountains. But when you introduce to the eagle the old dragon who has dwelt long among the rocks, and who creeps in and out of the caves of the earth, and set the two upon a hellish seat, then Pluto will blow strongly and drive out from the cold dragon a flying, fiery spirit whose great heat will burn up the feathers of the eagle and prepare a steam-bath so that the snow on the highest mountains must melt entirely and turn into water, whereby the mineral bath is rightly prepared and can give the king good fortune and health.[19]

This text almost seems the product of a slightly unbalanced writer, as it jumps from image to image with giddy abandon. Yet it too can be deciphered practically. The bridegroom's bath is a liquid prepared by the combat of the two fighters, also called eagle and dragon; these animals are shown on the combatants' swords in figure 6.4. Fortunately, Valentine mentions an eagle once again elsewhere in the book (probably an example of the dispersion of knowledge). There he equates that eagle with "salmiac," or sal ammoniac, a salt today called ammonium chloride.[20] One of ammonium chloride's characteristic properties is that it sublimes easily—that is, upon mild heating the salt vaporizes and then recondenses into a white salt in cooler parts of the flask. Given ammonium chloride's ability to sublime, eagle is an appropriate *Deckname* for it—both the salt and the bird fly through the air. (The modern term *volatilize* derives from the Latin *volare*, "to fly.") Accordingly, the "snow on the highest mountains" must refer to the deposit of pure white sal ammoniac that collects at the top of the vessel when the salt sublimes.

Identifying the dragon requires some knowledge of mineralogy. The fact that it lives in caves and around stones suggests saltpeter (potassium nitrate), a salt found naturally as a crystalline deposit on cave walls and in the stone foundations of stables. The remark that the dragon is "cold" further hints at saltpeter, for it tastes cool on the tongue, and it

Figure 6.4. The second key of Basil Valentine. From *Von dem grossen Stein der Uhralten* (Leipzig, 1602). By courtesy of the Department of Special Collections, Memorial Library, University of Wisconsin–Madison.

perceptibly lowers the temperature of water as it dissolves. Finally, "a flying, fiery spirit" can be driven out of saltpeter by heat—we call it nitric acid—which clinches the identification.

Replication proves the correctness of this interpretation. When ammonium chloride and potassium nitrate are mixed ("introduce to the eagle the old dragon"), placed in a retort in a furnace ("a hellish seat"), and heated strongly (Pluto, god of hell, starts to blow), a vigorous reaction (a fight) does ensue, and a highly corrosive acid distills over. This

Figure 6.5. The third key of Basil Valentine, encoding the major secret of his cohobation process. Note that an early reader recorded his own decipherment on the woodcut; he wrote the symbol for gold (☉) on the fox and the abbreviation for *amalgam* (aaa) next to the rooster. This interpretation differs from my own. From *Von dem grossen Stein der Uhralten* (Leipzig, 1602). By courtesy of the Department of Special Collections, Memorial Library, University of Wisconsin–Madison.

"mineral bath" is a type of *aqua regia*, an acid mixture capable of dissolving gold. The accompanying image shows Mercury positioned between the fighters and standing on wings. The sense here seems to be that the *medium between* the fighters is a winged Mercury—that is, a liquid that flies out from the fighting salts.

The text of the third key (fig. 6.5) describes how water conquers fire, and in the same way our fiery Sulfur must be prepared for this art and conquered with water . . . so that the king . . . is utterly shattered and made invisible. But his visible form must this time appear again.[21]

These allusive directions seem to describe the action of the prepared acid ("water") on the purified gold ("Sulfur"). Namely, the gold is dissolved ("utterly shattered") by the acid into a transparent solution ("made invisible"). Making "his visible form . . . appear again" implies that the gold reappears, suggesting that the solution should be evaporated to leave behind a residue, in this case gold chloride. Gold chloride is unstable in the presence of heat, so when its solution is evaporated, the residue decomposes quickly to produce gold once again—thus, the king's "visible form" reappears.

Valentine continues, "He who would prepare our unburnable Sulfur of all the Sages must take care to seek out our Sulfur in something where it is unburnable, which cannot be done unless the salty sea has swallowed the corpse, and then entirely spit it out once again."[22] The Sulfur of the Sages is a title for the Philosophers' Stone, goal of the twelve keys. To attain it, Valentine implies, one must use more acid (salty sea) to redissolve the gold (the corpse, that is, the residue from evaporating the first solution), and then distill it off again to restore ("spit out") the gold. This direction seems to make no sense chemically; it does not get us anywhere. Nevertheless, it describes a common chymical operation called *cobobation*, a technique not used by chemists today. In this process, a liquid is distilled off of some substance, and then the same liquid is poured over the residue and distilled off again—often for dozens of times in succession. Figure 6.6 illustrates the useless circularity of the process in the modern equivalent of an emblem. What could this repetition possibly achieve?

Valentine then reaches a new crescendo of the bizarre:

Then raise him up in degree so that he far surpasses all the other stars of heaven in brightness . . . this is the rose of our masters, scarlet in color, and the red dragon's blood. . . . Endow him with the flying power of a bird as much as he needs, thus the rooster will eat the fox, be drowned in water, be made living by fire, and be eaten in return by the fox, so that like and unlike are made alike.[23]

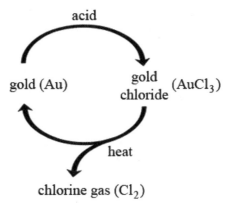

acid

gold (Au)

gold
chloride (AuCl$_3$)

heat

chlorine gas (Cl$_2$)

Figure 6.6. A modern "emblem" expressing the secret of the third key. Gold dissolves in acid, forming gold chloride; when the acid is distilled off, the gold chloride is decomposed by heat into gold and chlorine gas; the resultant gold is redissolved in acid, and so on.

A good laugh at Valentine's hodgepodge of weird imagery and high-flown language would not be inappropriate. Such language, however, typifies seventeenth-century chrysopoeia. The third key (fig. 6.5) shows the red dragon in the foreground, and that strangely carnivorous rooster both eating a fox and being eaten by him in the background. What connections familiar to early moderns could form the basis for the metaphors? Roosters had long been linked to the Sun (they crow at sunrise), and the Sun in turn to gold. Rooster would then be the *fourth Deckname* for "gold," previously encoded as king, Apollo, and Sulfur. The fox is a particular consumer of barnyard fowl (as in "a fox in the henhouse"), and consequently must be a new cover name for the acid that "eats" the gold. So, Valentine's allegory can be deciphered as follows: the gold drinks in the acid (rooster eats fox), is dissolved by it (drowns in water), reappears when heat evaporates the acid (rooster brought back to life by fire), and is then redissolved by fresh acid (fox eats rooster) during cohobation. This interpretation seems plausible—it both fits the text and is chemically possible—but the process still seems like running in place.

The direction, however, to give gold "flying power" that would "raise him up" above the stars, seems totally absurd. These phrases imply that the gold is somehow to be made volatile. It seems impossible to make something so heavy, solid, and stable in the fire as gold evaporate! In fact, the volatilization of gold represented both a desideratum and a point of ridicule in the early modern period. To "make the fixed volatile and the volatile fixed" was a guiding axiom for making the Philosophers' Stone, and few substances are more "fixed" (that is, nonvolatile) than gold. Making it volatile would therefore seem to be a huge step toward fulfilling an instruction handed down from the "ancient sages," and a clear sign that one was on the right path. At the same time, critics jeered at the notion as an example of foolish "alchemical fancy." In the 1717 comedy *Three Hours After Marriage*, for example, a character posing as a Polish alchemist brags of his skills in transmutation. When asked by a doctor how he succeeded, the sham alchemist rattles off a list of operations on gold, including its volatilization. The doctor, who has studied chrysopoeia, becomes suspicious at this point and warns him to "have a care what you assert. The volatilization of gold is not an obvious process. It is by great elegance of speech called, *fortitudo fortitudinis fortissima*."[24]

Despite being "the most difficult difficulty of the difficulty," in 1895, long after both the alchemical claim to volatilize gold and its ridicule had faded from memory, this very process—allegorically described three hundred years earlier by Basil Valentine—was actually rediscovered independently and chemically explained.[25] Valentine apparently *had* succeeded in volatilizing gold. Seventy years after him, so had Robert Boyle, who successfully deciphered and followed experimentally at least the first three of Valentine's *Twelve Keys* during his own quest to prepare the Philosophers' Stone.[26] I have tried the process myself and found it extremely difficult to carry out, but it eventually succeeded beautifully.[27]

Valentine's astonishing success depended on that seemingly useless cohobation. The repeated formation and decomposition of the gold chloride fills the distillation apparatus with chlorine gas. This toxic gas prevents the decomposition of the otherwise highly unstable gold chloride, allowing it to sublime as beautiful ruby-red crystals, or in Valentine's more vivid language, "the rose of our masters . . . and the red dragon's blood."

Four historical lessons can be drawn from this examination and replication of the first of the *Twelve Keys*. First, at least some cryptic texts and emblems dealing with making the Philosophers' Stone *do* encode real chemical processes that their authors carried out. Second, these bizarre allegories and emblems can be rationally and methodically deciphered, meaning that their authors constructed them carefully, not only to *con-*

ceal their knowledge, but also to *reveal* it in a measured way to the most talented and thus the most worthy readers. Third, readers *expected* such allegorical language and imagery to have a specific, discernible meaning; they labored at understanding it, and at least some of them succeeded well enough to replicate the processes.[28] Fourth, it is clear that at least some chrysopoeians had truly remarkable practical skills—Basil Valentine (whoever he really was) would be an acclaimed experimentalist even today. The volatilization of gold chloride is an extremely difficult and delicate operation with modern equipment—yet our would-be Benedictine monk accomplished this astonishing feat under the comparatively primitive working conditions (such as poor glass and charcoal fires) of the late sixteenth century.

Several subsequent keys continue to elude complete decipherment. One mysterious point is the identity of the queen/bride/Diana/Mercury that is eventually to be married to the king/bridegroom/Apollo/Sulfur/ rooster, that is, combined with the sublimed gold. Their marriage (union or combination) is depicted in the sixth key (fig. 6.7).

In 1618, the chrysopoetic author Michael Maier published a Latin translation of Valentine's book with elegant engravings replacing the original crude woodcuts.[29] Significantly, Maier inserted his own idea about the queen's identity by silently rearranging the first key (fig. 6.8; compare with 6.3). He shifted the wolf to the left in front of the king, and put Saturn on the right, now straddling a small oven, in front of the queen. These minor alterations to the engraving significantly change its meaning! Now the image depicts the purification of *both gold and silver*— gold purified with stibnite as before, but now also silver purified with lead, by a process known as cupellation. In cupellation, impure silver is melted with lead in a shallow dish made of bone ash, called a cupel. A stream of air is blown across the molten mixture, causing the lead and all the base metals alloyed with the silver to oxidize, whereupon they are either absorbed by the cupel or blown away. Pure silver is left behind. In Maier's engraving, Saturn is no longer the wolf's sire, but lead; the queen is no longer a yet-unidentified material to be combined in the future step with purified gold, but silver. Maier apparently believed he had correctly deciphered the meaning of the queen and decided to encode it into a reengraved emblem as a "gift" to his readers.

An early owner of the book from which figure 6.7 is reproduced came to a different conclusion, which he scribbled on the woodcut. Above the

Figure 6.7. The sixth key of Basil Valentine; the identity of the queen (and the bishop?) remain unclear, although an early reader inked in his own guesses. From *Von dem grossen Stein der Uhralten* (Leipzig, 1602). By courtesy of the Department of Special Collections, Memorial Library, University of Wisconsin–Madison.

queen's head he wrote the symbol for metallic antimony prepared using iron, the so-called martial regulus. These differing interpretations show how intelligent readers could draw conflicting conclusions from the same cryptic text and image (and I don't think that either Maier or the anonymous reader got it right).

The reader, however, probably has more justification for his view than does Maier, for Valentine himself might be giving hints in his ninth key regarding the queen's whereabouts. Here the text deals with colors and describes the stage in making the Philosophers' Stone in which the material sealed in the flask changes from black to white to red. The accompanying illustration (fig. 6.9) emblematizes the maturing stone. The king and queen are shown naked, while on their heads and feet rest the four birds symbolizing the successive stages in producing the stone—the black crow at the top, the multicolored peacock at the bottom, the white swan at the left, and the fiery red phoenix at the right. But if we step back a little and ignore the details, the overall design resolves into a circle surmounted by a cross formed by the oddly bent bodies of the king and queen: in short, the chymical symbol for antimony (stibnite).

As commonly happens in chrysopoetic texts, the end of Valentine's book is the most easily understood. The eleventh and twelfth keys describe operations after the stone has reached the red stage. In the last key, Valentine decides not employ "any philosophically flowery or figurative

Figure 6.8. Michael Maier's reengraved first key of Basil Valentine, published in *Tripus aureus* (Frankfurt, 1618). Courtesy of the Roy G. Neville Historical Chemical Library, Chemical Heritage Foundation, Philadelphia.

speech, but instead to reveal this key without any defect, in a complete and truthful process," and then gives a straightforward recipe for "fermenting" the stone by fusing it with gold. Despite the clear instruction, this combination of the Red Stone with gold is still depicted allegorically (fig. 6.10) as a lion eating a snake—probably a basilisk, one of the animals commonly linked to the Philosophers' Stone.[30]

Figure 6.9. The ninth key of Basil Valentine, showing the various colors of the maturing Philosophers' Stone, and perhaps hinting at a crucial ingredient.
From *Von dem grossen Stein der Uhralten* (Leipzig, 1602). By courtesy of the Department of Special Collections, Memorial Library, University of Wisconsin–Madison.

Figure 6.10. The twelfth key of Basil Valentine, depicting the "fermentation" of the Philosophers' Stone, enabling it to transmute base metals into gold. From *Von dem grossen Stein der Uhrhalten* (Leipzig, 1602). By courtesy of the Department of Special Collections, Memorial Library, University of Wisconsin–Madison.

Sources of Chrysopoetic Claims

There can now be no doubt that the exotic allegorical text and emblematic woodcuts in *Of the Great Stone* encode real chemistry that is groundbreaking for its time. How far could Valentine have gotten in his quest? Unless our current knowledge of chemistry is significantly flawed, he could not have succeeded in producing the transmutative Philosophers'

Stone. So where did the rest of the keys come from, if not from laboratory results? I would suggest that Valentine's book—and by extension many other chrysopoetic texts—results from the compilation of *three* different sources of information. Valentine's first keys are based on his own laboratory experience, and culminate in the volatilization of gold, a result so surprising and promising that it must have thrilled the author and convinced him that he was on track for producing the stone.[31] The middle keys, which are more obscure and less unambiguously decipherable, encode processes the author predicted as the next steps based on theoretical considerations, but which he had not yet accomplished. The final keys, the most straightforward of all, present material borrowed more or less directly from earlier books. Since the steps after sealing the prepared material in the "philosophical egg" had become virtually canonical by the end of the sixteenth century, there was no reason to encode them. Thus, by drawing on laboratory results, theoretical extrapolations, and textual precedents, Valentine mapped out a "plausible" route to the Philosophers' Stone, and encoded it as the *Twelve Keys*. His laboratory results, aligning beautifully with the axiom about making the fixed volatile, convinced him of his progress while the books of the "ancient sages" assured him that the goal had been reached by others before him. All that remained were the "missing links" in the middle of the process. Perhaps he was still hard at work on those operations when he wrote the *Twelve Keys*.

Although conjectural, this interpretation provides a plausible explanation of at least some chrysopoetic works. How many such texts can be explained by means of it remains an open question, but what I call "chymical optimism" based on encouraging laboratory results must play some role in texts claiming success with preparing the stone. Another example that corroborates this notion brings up one of the most remarkable characters of seventeenth-century chymistry—George Starkey.

Growing Your Own Gold: Starkey and the Philosophical Tree

Starkey represents a microcosm of seventeenth-century chymistry. He made and sold commercial products like perfumes, essential oils, and cosmetics; he labored to discover the alkahest; he practiced medicine and prepared new pharmaceuticals along Helmontian lines; he designed novel apparatus; he developed new theories of chymistry; he dabbled with animal palingenesis; he worked in refining and mining operations; he avidly explored metallic transmutation; and he strove to prepare the Philosophers' Stone. He wrote extensively, and (perhaps inadvertently) became one of the most highly regarded and widely read chrysopoetic authors of the seventeenth century.

George Starkey was born in Bermuda in 1628. His father, a Scottish minister who had relocated to the island, died while Starkey was young. As he showed considerable intellectual aptitude, his guardian sent him to be educated at a newly founded institution in the Massachusetts Bay Colony, Harvard College, where Starkey graduated in 1646. During his undergraduate days he developed a keen interest in chymistry and soon gained notoriety for his precocious knowledge and achievements.[32] In 1650, partly out of frustration at trying to conduct chymical experiments with inadequate materials and equipment, Starkey left America for England. He settled in London, and met several thinkers interested in new knowledge of all kinds, including an equally young (but vastly more wealthy) Robert Boyle. Starkey apparently encouraged Boyle's incipient interests in chymistry, and taught him a great deal from his own store of information and experience. At about the same time, Starkey began recounting remarkable stories about an adept he had met in America who possessed both the Red and the White Philosophers' Stones, and gave him a portion of the white one. This mysterious adept, who went by the name of Eirenaeus Philalethes (meaning "peaceful lover of truth"), also shared several of his manuscripts with Starkey. Starkey circulated some of these writings privately in London, where they generated great interest.

Although it seemed promising, Starkey's life turned out to be anything but easy. After establishing a lucrative practice as a physician, he eventually turned away his patients to devote himself full time to the pursuit of the "secrets of nature," namely chymistry in all its forms. But experimental research, then as now, was a costly and risky enterprise; Starkey soon found himself in debtor's prison. After his release, he renewed his search for better medicines and metallic transmutation, making and selling medicinal preparations, oils, and perfumes to help finance his experiments. Astonishingly, several laboratory notebooks he kept during the 1650s have survived to this day. These documents provide an extraordinary witness of the daily work and thoughts of a mid-seventeenth-century chymist in that they record his successes and his failures, his reflections on his projects and their progress, his method for proposing experiments based on the best theories of the day, and his use of experiments to amend those theories. The notebooks even record his methods for converting allegorical writings into practical laboratory instructions.[33]

When the Great Plague of London—the last major outbreak of bubonic plague in Europe—erupted in 1665, the licensed physicians fled the city. Starkey and his fellow advocates of chemiatria, however, stayed behind. They challenged the fleeing doctors to test whose medicines would cure more people of the disease. The challenge went unanswered. During the height of the contagion, Starkey caught the plague himself and died a few days later at the age of thirty-seven.

Although Starkey's life came to a premature end, the adept Eirenaeus Philalethes lived on. The Philalethes manuscripts Starkey had circulated began to be published and quickly gained enormous popularity. Among their careful readers was Sir Isaac Newton, who not only conducted experiments along the lines suggested by Philalethes but also adopted and further developed some of Philalethes' theories on the structure of matter.[34] Eager searches were made to locate more manuscripts as well as to find the mysterious adept himself. Fresh rumors about him and his whereabouts continued to emerge into the eighteenth century. Of course, these prized manuscripts were really Starkey's own compositions. One of his notebooks actually contains drafts of short and unfinished "Philalethes" treatises.[35] But he hid his authorship so well, and apparently told the tales of Philalethes' exploits so convincingly, that even his friend Boyle never knew the truth. Starkey published several books and pamphlets under his own name, but they never gained the renown of those by the famed Eirenaeus Philalethes. Starkey was keenly interested in developing coherent chymical theories, even while indulging in some of the most extravagant allegorical writing to be found anywhere in the whole literature of chrysopoeia. In his Philalethes writings, he recapitulates, categorizes, and critiques a range of contemporaneous ideas about the Philosophers' Stone and its preparation. These admirable qualities certainly contributed as much to the popularity of these texts as did exotic tales about their supposed author.

The approach of Philalethes, or rather of Starkey, to the problem of how to prepare the Philosophers' Stone is quite different from that of

Basil Valentine. The *Twelve Keys* exemplifies one major route to the stone, called the "wet way" (*via humida*) because it employs watery solvents—in this case, the acidic water of the fighters. Starkey exemplifies the other major route—the "dry way" (*via sicca*), which uses no such watery corrosives, only a "dry water," or as a stock phrase expresses it, "a water that does not wet the hands." Starkey's work falls under a school of chrysopoeia called *mercurialist*. For mercurialists, the key to achieving the Philosophers' Stone is preparing a Philosophical Mercury from ordinary mercury by a process of purification and "animation."[36] This "animation" refers not to anything like the experiments of Dr. Frankenstein but rather to the implantation of a "soul" into common mercury that corrects its usual coldness and wetness with an internal heat. This "soul" is not literally a spiritual entity any more than the "soul of gold" (*anima auri*) mentioned in chapter 5. Instead, it signifies a substance able to "heat" the mercury, giving it new properties. The term draws comparison with the animal soul that supplies "vital heat" to living creatures as long as the animal lives.

Interest in "animating" common mercury to provide a Philosophical Mercury is evident in dozens of texts from the sixteenth to the eighteenth century, and forms a coherent "research program" pursued by generations of hopeful mercurialists. Some chose to provide the heat of animation with gold, as in the case of Gaston Duclo in the late sixteenth century. Others chose various metals, quicklime, salts, or—like Starkey—antimony.[37] In fact, Starkey's preparation of "stellate regulus of antimony" (elemental antimony with a striking crystallization pattern on its surface; see plate 3) for use in this process beautifully exemplifies how chymical writers could use either highly allegorical or perfectly plain language, depending on their intended audience. The encoded style is appropriate for publications, where a veil of secrecy is obligatory to screen out those readers unworthy or potentially abusive of the secret. The straightforward style is used for private documents, where the readership is already restricted, as with Starkey's laboratory notebooks and his personal letters. This point bears underscoring; it has sometimes been said that "the alchemists" (as if they were all the same!) were incapable of clear expression, because either their ideas and processes had no clear meaning, or their language was in some way an "ecstatic" declaration rather than consciously encoded words. Such assertions are groundless.

In Philalethes' *Open Entrance to the Closed Palace of the King* (1667), the chapter titled "Of the first operation of the preparation of the sophic Mercury by the flying eagles" tells readers how to begin the animation process.

> Take our Fiery Dragon that hides the Magical Steel in its belly, four parts, of our Magnet, nine parts, mix them together with torrid Vulcan ... throw away the husk and take the kernel, purge thrice with fire and Sun, which will be easily done if Saturn sees his form in the mirror of Mars. Thence is made the Chamaeleon or our Chaos, in which all secrets are hidden in virtue not in act. This is the Hermaphroditical Infant infected with the biting of the Corascene mad dog. ... Yet there are two doves in the wood of Diana that assuage his mad rabies.[38]

Years before these obscure directions appeared in print, Starkey had given instructions for the same process in a letter to Boyle. In spring 1651, Starkey told his friend to take

> of antimony nine ounces, and of iron four ounces (which is the true proportion) ... let the fire be so strong as to cause the matter to flow ... poure it into a horne, and in the bottom wil be the Regulus, and a shining slag above it. Separate them when they are cold. ... You must have the mediation of Virgine Diana, that is, pure silver. ... Now Sir take of this Regulus one part, of pure silver two parts ...[39]

Elsewhere in the same letter, we find explanations for why this antimony regulus is called "hermaphroditic" and "rabid," and most of all, how to use it to prepare a Philosophical Mercury. For his part, Boyle himself found the results of this process so promising that he experimented with it for *nearly forty years* in attempts to turn the Mercury it produced into the Philosophers' Stone. He also allowed part of the letter to be copied (probably to Starkey's horror), and these copies found their way into numerous hands all across Europe. Newton himself owned one, but by the time it got to him it had passed through so many hands that its original connection to Starkey had been forgotten.[40]

Why was there so much interest in this Philosophical Mercury? Mercurialists maintained that it and common gold were the two starting materials for making the stone. Sealed in the philosophical egg, the two would react, display the necessary colors of black, white, and red, and produce the elixir. Many mercurialists (Starkey included) based their theory of the stone on the concept of *semina*, or seeds—those principles capable of organizing matter into specific substances and forms. They argued by analogy that since the seed of an apple, for example, is found only in an apple, the seed of gold must be found in gold alone. But experience teaches that simply mixing or melting gold with base metals does not cause transmutation; the seed of gold cannot act on other metals while it remains locked within the metallic body of gold—in that state it is dormant and weak. The Philosophical Mercury liberates and activates the seed of gold. The Mercury dissolves gold gently and naturally (rather than the violent, destructive way that acids act), freeing the seed with all its powers intact from the body of gold. The Mercury "nourishes" the seed and, upon extended heating in the flask, allows it to strengthen and multiply, finally culminating in the Philosophers' Stone, whose active principle is the highly activated seed of gold. The seed in the stone, no longer weak and inactive as it was while still in gold, can now bring about transmutation by reorganizing the underlying matter of a base metal into gold.

As mentioned previously, the word *seed* is metaphorical in this context; many who supported this theory of the stone's action stressed this point. In general, they did *not* see the metals as "alive" like a plant, or believe that metals were propagated by anything resembling the literal seeds a gardener puts into the ground. Yet the similitude drawn between the "seeds" of metals and the seeds of plants gave rise to an array of ancillary explanations and illustrations—which is the major purpose of metaphors, and why they remain crucial in the sciences to this day. Thus, mercurialist texts often draw on additional horticultural images related to notions of "seed." Just as ordinary water is necessary for the swelling and germination of ordinary seeds and their growth into plants aboveground, so too the Philosophical Mercury is considered a kind of "water" needed for the "germination" of gold and its "growth" into the Philosophers' Stone. Accordingly, Starkey's favored author, George Ripley, writes that with Philosophical Mercury,

> dyd Hermes moysture hys Tre:
> Wythyn hys Glas he made to grow upryght,
> Wyth Flowers dyscoloryd bewtyosely to syght.[41]

Jean Collesson, a mercurialist of the first half of the seventeenth century, writes that the value of Philosophical Mercury lies in its ability to "make gold vegetate and germinate." He assures his readers that if a prepared Mercury does not make gold "vegetate to sight," then it is not the true Mercury of the Philosophers.[42] Philalethes (or rather Starkey) also uses agricultural metaphors extensively, and catalogues their appearance in earlier authors' works. Noting the many authors who refer to plants and trees in the context of the Philosophers' Stone, Starkey writes that

> this tree of ours some have compared to one thing, and some to another; some to a Cypress or Fir-Tree, which indeed may seem to resemble it; others to Haw-Thorn Trees, as Ripley in his Gate of Cibation; others to Shrubs and Bushes, others to thick Woods.... I confess there is a similitude between our Germination and all these; ... others have called it their Coral, which is indeed the fittest comparison, for in our Tree there are Shoots and Sprigs, without any thing that may be properly likened to Leaves: as then Coral is an union of a Vegetable and a Stony nature, so is it in our Tree ...[43]

Within the context of the largely agrarian society of the early modern period, when people were closer to the experience of farming and gardening than most of us are today, botanical imagery provided an easily intelligible analogy for the conceptualization of the stone and the role of Philosophical Mercury in producing it. But can the vividness of the metaphor alone explain the long-term fascination with formulas for the Philosophical Mercury on the part of Starkey, Boyle, and so many others? Some writers with a dim view of chrysopoeia have attributed such continued interest to obsession—although obsession is usually a prerequisite for serious study and discovery—or some sort of delusion or "cycle of failure." Is there anything more behind it?

Once again, I thought that the best way to answer these questions was to re-create the experiments these authors "obsessed over" to see what they themselves saw. Starkey's laboratory notebooks provided a starting point. Unfortunately, only a few of his many notebooks survive, and so a complete plaintext description of his work did not exist. Working in part from surviving notebooks and correspondence, and filling the gaps with published interpretations of the Philalethes texts, I pieced together a reasonably complete account of his method for making and using Philosophical Mercury. As with Valentine's preparations, some of Starkey's procedures made no sense from a modern chemical perspective. But a little obsession goes a long way, and after a month of laborious and repetitive work I prepared a small quantity of what Starkey claimed to be the sought-after "animated" Philosophical Mercury, although it looked little different from what I had started with.

Following Starkey's hints, I mixed this Mercury with gold to produce a buttery mixture, which I placed in a flask approximating the form of a philosophical egg. The "egg" was sealed, buried in a sand bath, and heated (plate 4). For several weeks, I varied the heat, since the original texts do not (and in fact could not, in the age before laboratory thermometers) provide a clear indication of the temperature used originally. The mixture did little more during this time than swell slightly, increasing in fluidity and then becoming partly covered with warty excrescences. Finally, a few days after the right temperature had apparently been achieved, I arrived at the laboratory one morning to discover that the mixture had taken on a completely new—and extraordinarily surprising—appearance overnight. Only a gray amorphous mass lay at the bottom of the flask the day before, while a glittering and fully formed tree filled the vessel on the following morning (plates 5 and 6).

My first reaction to this sight was utter disbelief, and then—after becoming relatively certain that I had not taken leave of my senses—a sense of awe and wonder. Imagine, then, what any chrysopoeian of the seventeenth century must have thought when he witnessed such a sight. It certainly would have seemed a powerful vindication of his belief that the Philosophers' Mercury could liberate, activate, and nourish the "seed" of gold. It would have reminded him instantly of earlier authors who spoke of gold's "vegetation" and of the "Tree of Hermes." It would, in short, have been vivid and unquestionable proof that he had found the "entrance to the palace of the king," that is, the crucial threshold leading to the Philosophers' Stone. For the historian, the reality of this Philosophical Tree indicates unambiguously that at least some of the imagery of chrysopoeia, bizarre as it might seem, stems from the literal appearance of reacting chemicals.[44]

A surviving fragment of an otherwise lost Starkey notebook indicates clearly that the American alchemist saw the same Philosophical Tree.

On Tuesday, March 5, 1652, he recorded that a mixture of his Mercury and gold "stood 12 whole days for the greatest part of the time *in arborescentia*," that is, "in growing trees."[45] Given the results from replicating his process, we know now that he must be taken literally when he claims, "I have now in the fire several glasses of gold with that mercury which grow in the forme of trees."[46] In light of this visually impressive outcome, we can better understand the dogged pursuit of this route to the Philosophers' Stone; a sight such as this one must have provided enormous encouragement to continue experimenting.

Although other treelike (or dendritic) chemical "growths" were known in seventeenth-century chymistry, they are far different from the one shown in plate 6. The most well known was the "Tree of Diana," a simple crystallization of silver precipitated from a silver nitrate solution. These "growths" were familiar parlor tricks in the seventeenth century and persist today in the repertory of "chemical magic" shows.[47] In terms of both chemistry and historical significance, these trivial productions are not comparable to the jealously-guarded secret of the mercurialists' Philosophical Tree—tightly linked to chrysopoeia—which grows completely unexpectedly in a sealed vessel from an amorphous mixture of metals at high temperatures.

Starkey's continued experimentation apparently did not lead to the Philosophers' Stone; otherwise, it is doubtful he would have ended up in debtor's prison. The failure finally to obtain the stone, despite encouraging results such as gold's volatilization or its germination into a glittering tree, provokes the next question: why were so many people certain that the Philosophers' Stone could be, and had actually been, prepared? What evidence supported the widespread belief in its existence?

Evidence for the Philosophers' Stone

Nowadays, skepticism about the existence of the Philosophers' Stone is based primarily on the fact that its supposed powers run counter to accepted scientific matter theory. In the early modern period, however, the stone fit neatly into then-prevailing theories of matter. Transmutation was not contrary to contemporaneous systems of scientific thought. There existed no compelling theory with which to reject the stone's reality. On the contrary, various explanations for its powers, plausible in the context of the time, were available. Metallic transmutation appeared to occur spontaneously, albeit slowly, in nature; the chrysopoeian sought only a speedier means of effecting it, using what we might call (with some anachronism) a catalyst. The widespread tenet that all substances are composed of the same fundamental "stuff"—a view encapsulated in the ancient *ouroboros* (fig. 1.1) and reinvigorated by the most up-to-date ideas about matter in the seventeenth century—guaranteed at least the theoretical possibility of transforming anything into anything.

While these theoretical considerations rendered the stone a *possibility*, it took more to convince early modern people that it was also a *reality*. A second source of support came from eyewitness testimony. The literary heritage of alchemy contained descriptions of the stone and its effects dating back nearly a millennium. In the seventeenth century, a new genre of textual evidence emerged—the "transmutation history," testimonial accounts from recognized persons who had witnessed transmutation. These eyewitness accounts appeared both singly and as collections. One early example of the latter, published in 1604, is *Histories of Several Metallic Transmutations . . . for the Defense of Alchemy against the Madness of its Enemies* by Dutch author Ewald van Hoghelande. Such collections enjoyed a resurgence during the revival of chrysopoeia in late eighteenth-century Germany, and have even been compiled and published in our own era by alchemical believers and those who endeavor to vend "mysteries."[48]

Many accounts involve anonymous adepts who perform projection privately in front of aspiring chrysopoeians or skeptics. While some tales are so outlandish that they could scarcely fail to provoke a smirk from the modern reader (and probably had the same effect on the early modern reader), many are painstakingly precise, noting exact times, places, and persons in attendance, the quantity of gold or silver produced, the appearance of the transmuting agent (almost invariably a red powder), and so forth. Some of these demonstrations proved unhealthy for the demonstrator. News of transmutations performed in Berlin in 1701 by an apothecary's apprentice named Johann Friedrich Böttger (1682–1719) not only drew the mathematician and philosopher Gottfried Wilhelm Leibniz (1646–1716) to the scene but also led to Böttger's arrest by soldiers of Duke August the Strong of Saxony. Böttger spent the rest of his life in confinement, where, although he did not satisfy August's demands to make gold, he did help discover the secret to making porcelain, a commercial product that proved nearly as lucrative.[49] Such accounts—

Böttger was not the only transmuter imprisoned for his reputed knowledge—underscore a very practical reason for secrecy and anonymity in alchemy.

Some demonstrations took place more or less publicly, at a princely court or learned assembly, and were sometimes commemorated by the striking of coins or medallions from the transmuted metal itself.[50] In fact, by the end of the seventeenth century, a sufficient number of such coins had been struck that an entire dissertation was written about them, and many of these alchemical artifacts survive today (plate 7).[51]

The account published in 1667 by Johann Friedrich Helvetius (1625–1709), physician to the prince of Orange, achieved enormous notoriety.[52] On December 27, 1666, a stranger appeared at Helvetius's house in The Hague. Since Helvetius had previously written skeptically about chrysopoeia, his visitor engaged him in conversation on this topic. After some discussion, the stranger took out a small ivory box containing three heavy lumps of a glassy substance, which he claimed to be the Philosophers' Stone in sufficient quantity to produce twenty tons of gold. On a second visit, he gave Helvetius a piece of the stone "smaller than a rapeseed." After the man had left, Helvetius melted some lead, cast the particle of the stone on it according to the directions he had been given, and found his lead turned into gold.

The city mint master analyzed the metal and found it to be pure gold. Moreover, when he melted a sample with silver to assay it, the alchemical gold actually transmuted some of the added silver, yielding a 33-percent increase in the total amount of gold. As Helvetius suggests, this result was caused by an "excess of tincture."[53] Prominent intellectuals sought to verify the accounts—the philosopher Benedict Spinoza, for example, traveled to visit and question both Helvetius and the assayer.[54]

Another striking example was discovered recently among Robert Boyle's unpublished papers, held today at the Royal Society of London, the scientific institution he helped to found. Around 1680, Boyle wrote *Dialogue on the Transmutation and Melioration of Metals*, which argues in favor of the Philosophers' Stone and its powers, and includes several transmutation accounts that had been relayed to him privately. But this unpublished text also contains a vividly recounted first-person narrative of Boyle's own eyewitness of projective transmutation.[55] Boyle tells how he was introduced to a man who offered to show him an experiment that would transform lead into a mercury-like metallic liquid. Boyle sent his servant to obtain lead and crucibles for the experiment. When the experiment miscarried (the crucible fell over in the fire), the man offered to demonstrate another experiment, which Boyle mistakenly assumed would be a repetition of the miscarried one. He continues his account:

> The Lead being strongly melted, the Traveller opened a small piece of folded paper wherein there appear'd to be some grains, but not very many, of a powder that seemed somewhat transparent almost like exceeding small Rubies, and was of a very fine and beautifull red. Of this he tooke carelessly enough, and without weighing it, upon the point of a knife as much as I guessed to be about a grain or at most betwixt one grain and two, and then presenting me the haft of the knife he told me that I might if I pleas'd cast in the powder with my owne hand.[56]

But Boyle, who was often infirm, suffered from light-sensitive eyes such that he feared he would spill the powder accidentally while gazing into the glowing fire, and "therefore restoring the knife to the Traveller I desired him to cast in the powder himselfe which he did whilst I stood by and looked on."[57] After covering the crucible and heating it strongly for fifteen minutes, the two men took it out of the fire and let it cool. Boyle continues,

> The Crucible having been kept till it was cool enough to be managed without doeing harme we remov'd it to the window where, instead of running Mercury, I was surprised to find a solid Body, and my surprise was increased when the Crucible being inverted, though yett a little hott, the Mass that came out (and still retaind the figure of the lower part of the vessell) appear'd very yellow. And when I took it into my hand, it felt to my thinking manifestly heavier then so much Lead would have done. Upon this, turning my eyes with a somewhat amazed look upon the Traveller's face, he smiled and told me he thought I had sufficiently understood what kind of experiment that newly made was design'd to be.[58]

Boyle took the yellow metal with him; all his tests showed it to be gold. Soon thereafter, one of his friends, probably Edmund Dickinson (1624–1707), professor of medicine at Oxford and first physician to the king, told him of meeting the same traveler a few days later in Oxford. There Dickinson witnessed two transmutations—one starting with lead and the other with copper. "And in this last," writes Boyle, "the Physician [Dickinson] for fuller satisfaction would needs have the operation try'd on some of our English Copper farthings that he took out of his owne Pockett, which, though much more difficultly melted than the Lead had been, were no less really transmuted into Gold."[59]

This experience was enough for Boyle. He later told his confessor, the bishop Gilbert Burnet, that this incident gave him "convincing satisfaction" of the reality of the Philosophers' Stone and its ability to transmute metals.[60] In fact, in 1689 both Boyle and Burnet testified to the reality of transmutation before Parliament in order to have King Henry IV's 1404 law forbidding transmutation repealed; on the strength of their testimony, the old law was removed.[61]

Early moderns thus heard regular reports of successful transmutations being carried out all across Europe. These narratives provided them with continuing evidence for the real existence of the stone. Even if such accounts failed to convince every skeptic, they nevertheless offered fresh support and incentive to both practicing and armchair chrysopoeians. For them, the various sources of evidence reinforced one another. Contemporaneous testimonial accounts, affirmations by respected textual authorities, coherence with the best scientific theories of the day, and the sight of remarkable and striking laboratory phenomena such as the Philosophical Tree coalesced into a persuasive case that the Philosophers' Stone was both real and a worthwhile goal to pursue. Despite the debates over transmutation that had gone on for centuries, many leading intellectuals and natural philosophers (scientists, in our lingo) remained convinced of the reality of the stone and its powers. Many of the chymical practitioners who labored to uncover the secrets of chrysopoeia were serious thinkers and talented experimentalists, and several were also celebrated figures of the Scientific Revolution, such as Boyle and Newton.

This chapter focused primarily on revealing the actual practices and the chemistry that lies concealed beneath alchemy's forbidding tangle of secretive language and imagery. It was necessary for me to illustrate and stress such laboratory practices because of the widespread tendency to minimize chrysopoeia's rational and practical chemical content. But chymistry—despite its firm anchor in the practical transformation of material substances—had a much wider ambit than does modern chemistry. It was a rich tincture that colored multiple facets of early modern culture. It also offers modern readers a gateway to understanding early modern ways of thinking about and experiencing the world—ways that differ markedly from modern ones, and hold their own striking beauty and power. The next chapter approaches the subject of alchemy from other directions, sampling its broader context and vision.

SEVEN

THE WIDER WORLDS OF

CHYMISTRY

Chymistry was not a rarely glimpsed subject in the sixteenth and seventeenth centuries, nor did it exist in isolation from its intellectual and cultural context. Rather, it caught the attention and fired the imagination of many who never handled a crucible or an alembic. This chapter examines how alchemy permeated and interacted with early modern culture more broadly, well beyond the laboratories and workshops filled with the fumes of chymical fires. Equally important, the ways in which many chymists thought about their work and the world often differ from modern perspectives, and exemplify widespread ideas of the period. The study of alchemy in fact helps to illuminate broader aspects of the early modern worldview. Understanding alchemy requires being able to study it, at least for a time, through early modern eyes.

Alchemy's Disputed Place in Intellectual Culture

Allegorical images provide a good place to begin. Not every book of chymical emblems was created to be some sort of encoded laboratory notebook. Chymical books of emblems had multiple formats and multiple purposes. One of the most famous authors of these was Michael Maier (1568–1622).[1] His lavish *Atalanta fugiens* (*Atalanta Fleeing*) contains fifty beautiful engravings by the famous Swiss engraver Matthaeus Merian the Elder (1593–1650), and is the source for many of the alchemical images most commonly reproduced today. In contrast to Basil Valentine's organized sequence of "keys" that expound a single text and encode a single process, Maier's *Atalanta fugiens* is a florilegium of images. It collects imagery and expressions from an array of earlier authors—Hermes, Morienus, Valentine, and others—and assembles them into one of the most intricate and rich layerings of meaning to be found in chymistry.[2] Even though Maier probably did perform some laboratory work, his *Atalanta fugiens* lies much further from the world of laboratory practice than do the books of Valentine or George Starkey. (Some readers, including Sir Isaac Newton, nevertheless mined it for practical information about making the Philosophers' Stone.)

Each of the book's fifty chapters consists of five parts: a motto, an emblematic image, a six-line epigram (in both Latin and German), two pages of prose narrative, and, most innovatively, a piece of music arranged for three voices (fig. 7.1). The music provides the organizing theme: the tale of Atalanta and Hippomenes. In classical mythology, the athletic maiden Atalanta refused to give herself in marriage to anyone who could not defeat her in a footrace—and to make things interesting, she had the losers killed. Hippomenes took up Atalanta's challenge, but knowing that no man could outrun her, he ensured his victory using three golden apples acquired with the help of Aphrodite, goddess of love. When the race began and Atalanta darted ahead, Hippomenes rolled one of the apples past her. As she stopped to pick it up, he was able to take the lead. With the judicious use of his apples, Hippomenes won the race and Atalanta's hand.[3] In Maier's musical compositions, the soprano voice represents "Atalanta fleeing"; the tenor, "Hippomenes following"; and the bass, "the delaying apple."

Although the original sources of the imagery lie in earlier texts, Maier augments them with further connections, allusions, and meanings of his own. The epigrams are so intricate that it seems unlikely that any one reader would ever "get" all the references, allusions, connections, and puns.[4] Nor do we yet understand exactly how the music is connected to the images, but several theories have been offered.[5]

Figure 7.1. "The Earth is its nurse"; emblem 2 from Michael Maier, *Atalanta fugiens* (*Atalanta Fleeing*) (Oppenheim, 1618), pp. 16–17. The motto quotes the *Emerald Tablet*. Courtesy of the Roy G. Neville Historical Chemical Library, Chemical Heritage Foundation, Philadelphia.

16

FUGA II. in Quinta, infra.

Ein Säugmutter ist die Erden.

Atalanta Fugiens.

Romu lus hir ta lu pæ preſſiſſe ſed ubera capax

Jupiter & di ctus fer tur adeſſe fides.

Hippomenes Sequitur.

Romu lus hir ta lu pæ preſſiſſe ſed ubera capax

Jupiter & di ctus fer tur adeſſe fides.

Pomum Moratur.

Romulus hirta lupæ preſſiſſe ſed ubera capax

Jupiter & dictus fertur adeſſe fides.

II. Epigrammatis Latini verſio Germanica.

ROmulus von einer Wölffin iſt/ aber Jupiter geſäuget
Von einer Geiß/ wie wir ſagen/ daß der Weiſen Kinder nehret
Was Wunder iſt/ ſo wir ſagen/ daß der Weiſen Kinder nehret
Sey von der Erd/ ſo ſo ihre Milch hat gegeben?
So dann die Thier geſäuget han ſolche groſſe Helden gewiß/
Wie groß mag dann der ſeyn/ deſſen die Erd Säugmutter iſt?

EMBLE-

EMBLEMA II. De secretis Naturæ.

Nutrix ejus terra est.

EPIGRAMMA II.

ROmulus hirta prepresſiſſe, ſed ubera capræ
    Jupiter, & factis fertur adeſſe fides:
Quid mirum, teneræ SAPIENTUM viſcera PROLIS
    Si ferimus TERRAM lacte nutriſſe ſuo?
Parvulaſi tantas Heroas beſtia pavit,
    QUANTUS, qui NUTRIX TERRÆUS ORBIS, erit?
                               C   Apud

Atalanta fugiens represents Maier's endeavor to link chrysopoeia to the wider realm of intellectual and humanistic endeavors, and should be viewed as part of a much broader sixteenth-century humanist tradition of emblematics. This genre is exemplified most famously in the enormously popular work of Andrea Alciati (fig. 7.2), whose books of allegorical emblems were printed many times in the early modern period.[6] The basic presentation in both cases is a combination of motto-image-epigram, a grouping seen already in the *Rosarium philosophorum*. If the engraving in figure 7.2 were excised from its textual context, it might easily be taken for an emblem from a chrysopoetic text (note the *ouroboros*). But Alciati's text is certainly not chymical—it presents axioms of morality and virtue. Nonetheless, the imagery and format used by both Alciati and Maier, as well as by the many other crafters of emblems, are closely allied. Similarly, the audience both authors had in mind consisted principally of humanistically educated readers. Thus, the proliferation of chymical emblems in the seventeenth century must be regarded not only as a development within chymistry but also as part of a wider contemporaneous craze for emblematics of all sorts.

The popularity of emblematics partly depended on the early modern delight in "learned play," that is, piecing together and puzzling out meanings cleverly hidden beneath allusion and metaphor. Even popular periodicals of the seventeenth century, such as the Parisian monthly *Mercure galant*, contained "enigmas" in the form of both allegorical verses and pictorial emblems. The journal's editors encouraged readers to send in their interpretations, and published the best expositions in the following issue. Our closest modern equivalents are perhaps word jumbles, crosswords, sudoku puzzles, and other "brainteasers." Significantly, however, none of these modern forms employ the multivalent power of pictorial imagery or encode wise, moral, or learned messages, which together represent the lifeblood of the early modern genre. All the same, Maier's work is, at the simplest level, an early seventeenth-century book of brainteasers for the learned. Its title page advertises *Atalanta fugiens* as "accommodated partly to the eyes and the intellect . . . partly to the ears and the recreation of the mind," thus casting it as a book of erudite puzzles and intellectual delight.

Maier's main goal for *Atalanta fugiens*, however, lies higher. An accomplished humanist and poet, he uses the poetry, music, learned play, and

NEPTVNI *tubicen (cuius pars vltima cetum*
  Aequoreum facies indicat eſſe deum)
Serpentis medio Triton comprenditur orbe,
  Qui caudam inſerto mordicus ore tenet.
Fama viros animo inſignes, præclaraꝗ geſta
  Proſequitur; toto mandat & orbe legi.

Figure 7.2. "Acquire immortality from the study of literature"; emblem 132 from
Andreae Alciati emblemata (Antwerp, 1577), p. 449. Courtesy of Johns Hopkins
University, Sheridan Libraries, Rare Books and Manuscripts Department.

beautiful images of his book to link chymistry to the liberal and the fine arts. His purpose, then, is not simply to entertain readers but rather to ennoble a practice generally considered dirty and laborious by making it attractive to humanist contemporaries. Maier tells his readers that

> in this life, the more anyone approaches a divine nature, the more he rejoices and delights in things to be investigated by the intellect, things that are subtle, marvelous, and rare. . . . For the purpose of cultivating our intellect, God has hidden infinite secrets in nature . . . chymical secrets are not the least of these, but are, after the investigation of divine things, the first and most precious of all.[7]

In other words, intellectuals should pay attention to chymistry. A keen knowledge of classical literature and history, mythology, mathematics, poetry, astronomy, music, theology, and of course chymistry are all essential prerequisites to reading, viewing, hearing—and perhaps rarest of all, *enjoying—Atalanta fugiens*. The greater the reader's breadth of knowledge, the greater his understanding; the greater his understanding, the greater his delight. Moreover, the hunt for hidden connections and meanings in *Atalanta fugiens* parallels that for the secrets divinely hidden in the natural world—a search that, Maier argues, can be carried out especially well using chymistry.

Atalanta fugiens is one instance of continuing attempts to address chymistry's shaky cultural and intellectual position—an issue that plagued chymists from the Middle Ages through the eighteenth century. As a "mixed" discipline of head and hand, of elevated ideas and laborious work, of promise and failure, pursued by practitioners from every rung of the social and intellectual ladders, it was (and remains to some extent) difficult to pin down. Accordingly, perhaps the most constant feature of early modern chymistry is its stubbornly bifurcated reception and ambiguous reputation. It was both condemned as fraudulent or useless and praised as powerful, even sacred, in almost every context.

As noted earlier, alchemy failed to find a foothold in the medieval university. It fared no better with Renaissance humanists, who endeavored to establish new intellectual circles outside university culture. The early exponents of humanism tended to condemn it.[8] At the start of the fourteenth century, the poet Dante (1265–1321) placed alchemists deep in the eighth circle of Hell, alongside counterfeiters and forgers. During his infernal tour in *The Divine Comedy*, he meets the soul of a man he knew in life who had been executed in 1293. The tortured soul tells him, "I am the shade of Capocchio, who falsified metals with alchemy. And you should remember . . . how I was a good ape of nature."[9] Dante's emphasis here rests on the immorality of making things seem what they are not—thus his linkage of alchemy to counterfeiting and forgery, as seen also in Pope John XXII's decretal, which was promulgated during Dante's lifetime. The condemned souls in *The Divine Comedy* merely aped nature clumsily, making poor, even ridiculous, likenesses of its productions instead of following and imitating it even to the point of exceeding its products, as Roger Bacon had claimed alchemy was able to do. Just a little later, Petrarch (1304–1374), in his *Remedies for Fortunes Fair and Foul*, decries chrysopoeia equally one-sidedly as an empty and worthless practice whose only successful productions are "smoke, ashes, sweat, sighs, words, trickery, and degradation."[10]

Later humanists, interested primarily in elegant language and classical texts (neither of which chymistry could boast), generally followed suit. By the sixteenth century, however, some of these scholars took on the challenge of "humanizing" previously neglected or spurned fields of knowledge and practice. The work of Georgius Agricola (1494–1555) provides a pertinent example. A humanist trained in Latin and Greek, and later a physician practicing in mining towns, Agricola tried to draw mining and metallurgy into learned circles. His massive *De re metallica* (*On Metallic Matter*) encyclopedically describes the finding, digging, and working of mines as well as the smelting and refining of mineral products. Agricola was not trying to create a miner's handbook (in fact, his descriptions are often technically inaccurate) but rather to systematize and ennoble mining by dressing it up as a learned, humanistic discipline. Accordingly, his book replaces barbarous Germanic mining terms with an erudite Greco-Latin vocabulary. It gives mining a classical pedigree by frequently referencing Greek and Roman authors, and its large and artistic illustrations render the volume pleasing to the eye.[11] The lavish format (and resultant high price) of *De re metallica* signals its privileged audience. A miner, assayer, or engineer was not meant to own a copy for reference any more than a practicing chrysopoeian was meant to have *Atalanta fugiens* lying open next to his furnace.

Alchemy received analogous treatment at the hands of Giovanni Aurelio Augurello (1441–1524). A humanist poet and admirer of Petrarch, Augurello published a lengthy poem titled *Chrysopoeia* in 1515.[12] The poem imitates the style of Virgil's *Georgics*, in which the Roman poet gives farming the polish of elegant Latin verse. Augurello likewise dresses up alchemy with classical language, literary style, and learned allusions. He dedicated his work to Pope Leo X, himself a noted humanist, who was said (with little evidence) to have sent the poet an empty purse as recompense for his efforts, with the implication that Augurello, given his knowledge of chrysopoeia, could fill it himself. Augurello's mastery of chrysopoetic concepts indicates that he must have immersed himself in the relevant literature, although it is improbable that he ever dirtied his hands with crucibles. *Chrysopoeia* enjoyed wide popularity, and eventually became a source for chymists who, in their quest for the Philosophers' Stone, combed it and paraphrased it for hints about how to proceed practically.

One of Augurello's techniques for giving chrysopoeia a classical pedigree was to interpret Greek and Roman mythology as veiled descriptions of chymical processes. Jason and the Argonauts' search for the Golden Fleece thus became an allegory for the search for transmutation. The labors of Hercules and the loves of Venus likewise were interpreted as containing veiled chymical information. The reading of classical mythology as chymical allegory developed into a standard part of chymical literature. It appears, for example, in *Atalanta fugiens* but even more so in Maier's *Secrets Most Secret* (*Arcana arcanissima*).[13] Some chrysopoeians even argued that the chymical interpretation of mythology was the only way to save the ancients from the charge of blasphemy against their gods, given the extremely unflattering view of divinities that a literal reading of mythology would provide.

The proliferation of such pedigree building and allegorical readings eventually backfired. As the early modern period wore on, chrysopoeians began reading virtually *everything* as chymical allegory, and co-opting an astonishing range of ancients as adepts. This exuberance extended not only to Homer, Ovid, and other classical authors but also to medieval epics and the Bible. It was, of course, the last of these that provoked the strongest reactions from both Catholic authors concerned about unorthodox readings of scripture and Protestant ones committed to more literal readings.[14] Thomas Sprat assailed the practice in his 1667 *History of the Royal Society*: "This secret [of the Philosophers' Stone] they prosecute so impetuously, that they believe they see some footsteps of it, in every line of *Moses, Solomon,* or *Virgil*."[15] Herman Boerhaave (1668–1738), in giving his inaugural oration at the University of Leiden in 1718, expressed shame at "impious" chymists: "How I wish . . . these raving men had restrained themselves and not wished to interpret the Sacred Scriptures in terms of chymical principles and elements!"[16] If biblical passages could be interpreted as providing directions for laboratory work, biblical personages must have been practicing chrysopoeians. When Moses burned the golden calf and made the Israelites drink it, was he not preparing potable gold using knowledge acquired in Egypt? Solomon's great wisdom must have extended to transmutation; therefore, the gold that supposedly came from far-off Ophir must actually have been produced using the Philosophers' Stone.[17]

The addition of biblical patriarchs and ancient pagans to chymistry's lineage gave the discipline (and its practitioners) an ancient pedigree and status that otherwise was lacking.[18] Although the co-opting of ancient religious figures—Noah, Moses, St. John the Apostle, and others—as chymical adepts first appeared in the Latin Middle Ages, and the naming of Hermes Trismegestus as alchemy's founder in the Islamic period, the early modern period went further.[19] Through the seventeenth century, various histories of chymistry pushed its origins further and further back in time, and co-opted increasing numbers of ancients—both biblical and pagan—as adepts. Chymistry itself became one part of a broader "Hermetic knowledge," traceable not just to Hermes but to the most distant and venerable past, to a body of knowledge known as the "original wisdom" (*prisca sapientia*) revealed by God to the ancient patriarchs—in some versions to Adam himself—and passed down from generation to generation.[20] Unfortunately, this knowledge gradually eroded along the way, corrupted by successive transmission; pagan mythology was actually a degraded, misunderstood version of that original knowledge, hence the possibility of and need for interpreting it. While this lengthening pedigree and expanding ambit aimed largely at improving chymistry's standing, it also promoted chymical interpretations of increasingly far-flung materials, which in turn invited fresh ridicule from critics.

Chymistry in Literature and Art

Poets, painters, and playwrights discovered chymistry as well, making use of its ideas and images as well as its schizophrenic reputation. In so doing they left records of how the Noble Art was perceived (and adapted) by its contemporaries, and testimony of how familiar chymistry's basic operations and ideas had become. Their creations help flesh out a portrait of chymistry on a broader cultural scale.

Geoffrey Chaucer (circa 1343–1400) takes a position in his *Canterbury Tales* that is more nuanced than either Dante's or Petrarch's. The "Canon's Yeoman's Tale" recounts endless failed experiments leading to bankruptcy and illness, and tells of a deceitful canon who defrauds a priest with sleight of hand and a false powder of projection. Yet Chaucer does not thereby conclude that transmutational alchemy is false; rather, it is a privileged kind of knowledge with which only a select few should dare to meddle.

> Let no man trouble to explore this art
> If he can't understand the aims and jargon
> Of alchemists—and if he does, then
> He is a pretty foolish sort of man
> Because this art and science is, said he,
> Indeed a mystery in a mystery.
>
> And so I conclude: since God in heaven
> Will not permit alchemists to explain
> How anyone may discover this stone,
> My best advice is this—let it alone.[21]

Chaucer's tale warns rather than condemns. Most aspirants will fail to prepare the Philosophers' Stone, unless they have deep understanding of alchemy's particular "aims and jargon." The tale also reveals Chaucer's own familiarity with medieval alchemical authors and texts; he cites the pseudo-Arnald and other authorities, and paraphrases al-Rāzī. Some later chrysopoeians even looked on Chaucer as an adept.[22]

Chymical authors themselves frequently offered advice akin to Chaucer's, dissuading the wrong sort of readers from undertaking the work. The fifteenth-century chrysopoeian Thomas Norton, for example, lists the kinds of people who are doomed to failure because of insufficient leisure, learning, funds, or intelligence, and concludes,

> For trewly he that is not a greate Clerke
> Is nice and lewde to medle with this warke.
> ... For it is a most profound Philosophie
> The subtill science of holy Alkimy[23]

The visual arts repeat such warnings. In fact, the chymist became a stock figure in the Netherlandish art of the sixteenth and seventeenth centuries. Dutch and Flemish artists produced thousands of paintings depicting chymists. While many such paintings are full of accurately rendered apparatus of glass, metal, ceramic, and stone, they were not intended to be "photographic" representations.[24] Their primary purpose was to offer a moral lesson—often a slightly ambiguous one—that had to be teased out by the viewer, not unlike contemporaneous emblems.

Among the earliest of these images is Pieter Brueghel's mid-sixteenth-century drawing titled *Alghemist*. Thanks to a print produced by Philips Galle in 1558 (fig. 7.3), Brueghel's composition achieved wide dissemination, and proved enormously influential. The scene shows a family's ruin. At the center of the composition, a distraught woman holding an empty moneybag gestures behind her chymist-husband, who is dropping their last coin into his crucible. A fool squatting on the floor provides silent commentary by mimicking the chymist. The children, meanwhile, play in empty cupboards, underscoring the poverty caused by their father's hopeful schemes. In the background, the whole family—the eldest child still with an empty cauldron stuck on his head—suffers the consequences of the father's labors by going into the poorhouse. The meaning of the seated scholar on the right is slightly ambiguous. He might be reading out instructions to the practitioner, and yet he seems to exist apart from everyone else as a kind of a commentator, gesturing to the viewer to behold life gone awry. In Brueghel's original, the scholar points to the words *al-ghemist* in large print in his book—a pun on a Dutch phrase for "everything is lost."[25] In Galle's print, an added motto steers the drawing's meaning toward warning rather than condemnation. Its verses partly parody the style of the *Emerald Tablet* and other chymical axioms and begin, "The ignorant ought to put up with things and then labor diligently." It thus seems to say—like Chaucer's Canon's Yeoman—that

Figure 7.3. Philips Galle, *Alghemist*, 1558; engraving after a drawing
by Pieter Brueghel the Elder.

alchemy is not for everyone; it is not the means, and certainly not a shortcut, for improving one's lot in life. The less-gifted should steer clear of it,
and work diligently at something they are better suited for and something more certain of success.

Brueghel's image of a family ruined by chymistry spawned a dynasty
of variations on the theme. Dozens of later prints and paintings borrow
elements from his *Alghemist*. Adriaen van de Venne's 1636 *Rijcke-armoede*,
or *Rich Poverty* (plate 8), shows the father working busily at the furnace,
completely oblivious to the plight of his needy family. His wife looks
to heaven and displays the family's last coin in her outstretched hand
while the children beg for sustenance. Richard Brakenburgh reprises
many of these same features in his own larger-scale presentation of the
theme (plate 9). In it the hopeful chymist points to his wrapped powders, as if to say to his wife, "This time it's *really* going to work!" She
instead gestures toward their youngest son, uselessly pumping air at his

own upended crucible, not only wasting time and charcoal (both costly commodities) but also taking the place of the fool in Brueghel's drawing—an allusion obvious to anyone familiar with both artworks. Besides failing in his duty to support his family, the father has also corrupted his progeny by his poor example. Indeed, an older son, standing behind the father, pumps merrily on the bellows lever, joining in his father's wasteful activity. The common message is that the pursuit of wealth by means of chrysopoeia brings familial ruin. The moral in these paintings seems akin to the ancient proverb "Shoemaker, stick to your last."

David Teniers the Younger (1610–1690) was the most prolific creator of alchemist genre paintings. Interestingly, he portrays the chymist in quite a different light. In several paintings, Teniers places a full moneybag dangling from the chymist's belt, possibly as a direct refutation of Brueghel and his followers (plate 10). There is no ruined family, no starving children, no trace of foolery or impending disaster. The chymist, despite the messiness of his workshop, displays industry and productivity. Late in life, Teniers even painted a self-portrait depicting himself as a chymist, perhaps intending to underline a common theme of creativity and production for both painter and chymist—the painter, who combines simple materials to produce precious art, and the chymist, who combines simple substances to produce more valuable ones.

Like Teniers, the slightly younger Thomas Wijck (1616–1677) also produced paintings in which the chymist is a figure of good morals rather than bad ones. In an intimate portrait-like scene (plate 11), Wijck's chymist bears all the attributes of a scholar. Well-dressed, he sits reading amid books and papers while minding his distillation. A string of letters hangs by the window. Rather than chaos and ruin, the scene is one of peace and tranquility. In another painting, again possibly conceived as a conscious counterpoint to the Brueghel-type images, Wijck depicts a scene of almost amusing domestic harmony (plate 12). While the mother and children prepare dinner, the father works in his study, one of his distilling alembics sharing the stove with the family's meal.

Taken together, these artworks reiterate and comment on the ambiguous valuation of chymistry in early modern society. They depict it both as an obsessive pursuit that can lead unwary or unwise followers to ruin, poverty, and immorality, and as a productive labor, requiring steady industry, learning, and skill.

Seventeenth-century theatre, however, nearly always portrays the chymist comically. Sometimes he is a mere bungler, other times an outright fraud or charlatan. Ben Jonson's *The Alchemist* (1610) is the best-known example.[26] Its main character, Subtle, is a fast-talking con artist whose aim is to obtain gold from the baser instincts of greedy and unwise patrons rather than from baser metals. While he promises a completed (but ever-delayed) Philosophers' Stone, he cajoles gifts and lucre from those being

Figure 7.4. A scene from act 3 of *Les souffleurs*. Note the various smoking furnaces and pieces of apparatus, especially the large furnace at the back with a philosophical egg (containing the maturing Philosophers' Stone) perched on top. Chymical apparatus and bellows are used as trophy decorations on the rear wall. From *Les Souffleurs; ou, La pierre philosophale d'Arlequin* (Amsterdam, 1695); collection of the author.

gulled. Jonson also wrote masques—court spectacles that might well be called emblematics in performance. Several of them, produced at about the same time as *The Alchemist*, use alchemical imagery and concepts. His *Mercury Vindicated from the Alchemists at Court* borrows heavily from a contemporaneous work by Polish chymist Michael Sendivogius, and the *Alchemist* displays Jonson's significant familiarity with the relevant terminology and expressions, even when they are being parodied.[27]

In William Congreve's (1637–1708) *The Old Batchelor*, one character is urged to extort money from another, with the advice that "a little of thy Chymistry Tom, may extract Gold from that Dirt." "Faith," his companion responds, "I am as poor as a Chymist, and would be as industrious," thereby once again combining the contrasting themes of poverty and diligent labor found in contemporaneous genre paintings. A similar artful dissonance reappears in Congreve's *Way of the World*, which describes a lovesick woman as having "a Heart full of Hope, and a Head full of Care, like any Chymist upon the Day of Projection."[28] References such as these to chymists and chymistry indicate how familiar audiences must have been with the character of the chymist and the outlines of his craft.

Satire and chymically tinged humor appear also in *Les Souffleurs*, a comedy written in 1694 for the Parisian Théâtre Italien. The title translates as "The Puffers," a term applied pejoratively to overly hopeful chymists in reference to their constant blowing (*souffler*) at the coals under their crucibles. (As an alternate translation, I might suggest "The Blowhards.") *Les Souffleurs* presents a pair of neighbors trying to make the Philosophers' Stone, all the while oblivious to the standard amorous intrigues going on under their noses. The dialogue is peppered with chymical puns and allusions. When the stone is almost complete, a motley band of likeminded persons gathers to witness the projection and to join in a song, replete with chorus, extolling the virtues of their art (fig. 7.4).

> How wonderful is chymistry!
> With its prodigious effects
> It makes us equal to the gods
> Through the Elixir and potable gold.
>
> The most contemptible poverty,
> The old age that overcomes us,
> The least curable illness,

And even the inexorable Fates,
All feel the miraculous effect
Of our incomparable Stone.

How wonderful is the chymical art!
How marvelous its power![29]

Chymistry in Poetry

While playwrights parodied the chymist, poets used chymical themes and concepts both positively and negatively. William Shakespeare (1564–1616) elegantly draws on the transformative power central to chymistry in his thirty-third sonnet:

> Full many a glorious morning have I seen
> Flatter the mountain-tops with sovereign eye,
> Kissing with golden face the meadows green,
> Gilding pale streams with heavenly alchemy.

Around the same time, John Donne (1572–1631) used the equally constant hope and failure of chrysopoeians to exemplify the exaggerated optimism and ultimately unfulfilled hopes of a new bridegroom.

> Oh! 'tis imposture all:
> And as no chymique yet th'Elixar got,
> But glorifies his pregnant pot
> If by the way to him befall
> Some odoriferous thing, or medicinall,
> So, lovers dreame a rich and long delight,
> But get a winter-seeming summer's night.[30]

The Image of the "False Alchemist"

The bifurcated portrayal of chymists in art, literature, and theatre surely played a role in setting the foundations for the eighteenth-century division of "alchemy" and "chemistry" detailed in chapter 4. The images and stories of failure and frauds helped create a stereotyped category into which all hopeful chrysopoeians were eventually cast, one that proved remarkably durable. Accordingly, the well-worn category of alchemical frauds deserves direct attention.

Descriptions of frauds perpetrated by sham chrysopoeians form a continuous tradition beginning no later than the Islamic Middle Ages and culminating in the moral attacks on chrysopoeia in the eighteenth century. Such accounts were composed not only by critics and satirists, however, but also by chrysopoeians themselves. The latter wanted to warn the unwary of possible sleights of hand, and also to distinguish themselves clearly from the disreputable stereotypes that had become notorious in literature and the public mind.[31] There are also many stories about chymists executed for defrauding powerful patrons with golden promises they could not keep. Many of these accounts are true, but it would be wrong to consider all such unfortunates true frauds in the modern sense of the word. In many cases, these were practitioners who had acquired or developed some process for improving metals—not always involving anything as grand as the Philosophers' Stone—or enhancing the efficiency of a mining or refining operation. Confident of their future success, perhaps after promising signs on a small scale, they signed legal contracts with princely patrons. These contract-holding alchemists have recently (and felicitously) been called entrepreneurial chymists."[32] Their contracts stipulated how much the patron would provide in lodging, workspace, and materials, and established specific terms for deliverables and delivery dates. When the process failed, the practitioner failed to fulfill his contract. Under the usual understanding of contractual obligations—at least in the German lands, where most such contracts were made—this failure was considered *Betrügerei*, a word usually translated as "fraud." But these practitioners were not necessarily dishonest. This category of crime referred generally to promising anything that could not be delivered—in short, defrauding the ruler—and such a crime was punishable by death. The execution of failed chymists is a predominantly German phenomenon; we have but few records of such executions in France or England. This difference is probably more attributable to different systems of law than to different practices or practitioners.

The entrepreneurial chymists typically were not writers of learned treatises. In fact, well-informed authors tended to criticize them as pretenders, drones, and false chymists—or, often enough, frauds. At best, these authors characterized the entrepreneurial chymists as "process-mongers" who lacked a sound philosophical or theoretical basis for what they were attempting. To some extent this categorization was accurate, even if the consequent moral evaluation remains dubious. It is certainly correct to differentiate between the developed theorizations and experimental programs of someone like Starkey or Valentine and the more empirical efforts by those who signed contracts with rulers. Yet both groups represent important facets of early modern chymistry.[33] Those who sweated over smoky furnaces and swapped recipes but did not publish books were certainly more numerous than those who did publish, and possibly more visible in their own day and thus more responsible for generating popular impressions of the chymist, even if they did not participate as directly in the intellectual developments of the discipline.

Princely and royal courts were centers of patronage for a wider array of chymists than simply the entrepreneurial alchemists. In France, the court of Henri IV (reigned 1589–1610) buzzed with promoters of the new Paracelsian medicine, who saw both their novel therapeutics and their monarch as symbols of a new age. Distillation houses for the chymical production of medicinal waters operated from the grand imperial palace of Spain's El Escorial, to Francesco and Cosimo de' Medici's Florence, to numerous minor noble courts throughout German lands. Workshops for the improvement of mineral and metal resources were also a common fixture. Moritz "the Learned" of Hessen-Kassel (1572–1632) not only founded the first university professorship in chymical medicine but also presided over a constellation of competing court chymists, active in both chrysopoeia and chemiatria. Courts of several Holy Roman emperors in Prague and Vienna attracted chymists throughout the seventeenth century, and numerous "public demonstrations" of transmutation are supposed to have been carried out there. In short, chymistry—in all its dimensions—was not something restricted to solitary or isolated laboratories or private studies; it was also involved, often with a high profile, in early modern court culture.[34]

Alchemy and Religious Literature

Like the authors of secular literature, religious writers and orators drew on chymical ideas.[35] A natural affinity certainly exists between religious and moral ideas and the themes of purification and improvement found everywhere in chymistry. The Bible contains repeated comparisons between the refining of precious metals by fire and the testing and purification of the human heart.[36] Martin Luther (1483–1546) praised chymistry for its allegorical exemplification of Christian principles, even while he remained skeptical of chrysopoeia.[37] The central chymical operation of distillation—the separation of a pure, volatile (that is, "spiritual") substance from the crasser, baser components of a mixture—appears frequently as a trope in devotional literature. For example, the bishop Jean-Pierre Camus (1584–1652) provides a "laboratory formula" for carrying out a "spiritual alchemy":

> Let us put all our good and bad thoughts, affections, passions, vices, and virtues all mixed together into the alembic of our understanding. Place it then upon the memory and recollection of the eternal fire as if upon a furnace, and we shall see some marvelous subtle effects. This fiery cogitation will separate the confused elements, the hullabaloo of ambition, the earth of greed and lust, the winds of vanity, the waters of covetousness, the air of presumptions. It will dissipate all these follies, destroy the dregs and lees of a thousand earthly desires, in order to extract beautiful and completely heavenly conceptions from them . . . it will dissolve all our vices and sins, and extract from our souls a quintessence of piety and devotion. . . . Now then, isn't this a fine chymistry?[38]

Virtually every aspect of chymistry—spagyria, Paracelsian medicine, chrysopoeia, and a host of productive chymical operations—appears in innumerable sermons and tracts both Catholic and Protestant. Religious writers freely borrowed ideas and images from transmutational and medical chymistry for use as metaphors—just as writers on transmutation freely borrowed ideas and images from religion and theology for their own metaphorical purposes.[39] St. Francis de Sales (1567–1622), for example, in writing of the transformative power of love, exclaims, "O holy and sacred alchemy! O divine powder of projection whereby all the metals of our passions, affections, and actions are converted into the purest gold of heavenly love." Another orator of the same period calls divine grace "the true Philosophers' Stone that changes everything into gold."[40]

The use of chymical ideas as a source of rhetorical ornament or metaphor in religion (and vice versa) is straightforward and easy to

understand. But the interrelationships of chymistry and religion are far
deeper and more complex. Every chapter in this book has touched on
the alchemy-religion dynamic to one extent or another. These interrelationships are crucial—not only for gaining a fuller understanding of
chymistry in particular but also for their ability to illustrate larger points
of profound importance about early modern outlooks and worldviews
more generally. One place to start unraveling this extremely complex
skein is with the chymists' repeated claims of the divine origin and status
of their arcane knowledge.

Alchemy as the "Gift of God"

Chymists often called the knowledge of how to prepare the Philosophers'
Stone or other grand chymical secrets a *donum dei*—a gift of God. Thomas
Norton, for example, begins his *Ordinall of Alchimy* by declaring,

> Maistryefull merveylous and Archimastrye
> Is the tincture of holi Alkimy:
> A wonderfull Science, secret Philosophie,
> A singular Grace & gifte of th'almightie.[41]

Both the earliest surviving manuscript of the work and its printed
edition contain an illustration depicting a student on his knees receiv-
ing the secrets of alchemy from his master (fig. 7.5).[42] The seated master
tells the student, "Receive the gift of God under a sacred seal," while
the student responds, "I will keep the secrets of holy alchemy secretly."
Hovering over the scene is the Holy Spirit depicted as a dove, flanked by
angels carrying banners inscribed with verses from the Psalms (Psalm
45:7 and 27:14), thus giving a powerful sense of divine revelation to the
scene. Because most people today tend not to call any body of natural
knowledge holy or a a gift of God, such expressions and images seem to
mark out chrysopoeia as something special, something of a significantly
different order from other knowledge, something more akin to religious
knowledge than to natural knowledge. To be sure, such claims for its
divine origin and holy character have been used to support nineteenth-
and twentieth-century notions that alchemy is at root a spiritual, supernatural, or religious practice. But these expressions must be historically
contextualized to be understood as their authors intended.

Figure 7.5. The student receives the secrets of alchemy from his master, promising
to preserve them in secrecy. From Thomas Norton, *Ordinall of Alchimy*; engraving
in *Theatrum chemicum britannicum* (London, 1652). Courtesy of the Roy G. Neville
Historical Chemical Library, Chemical Heritage Foundation, Philadelphia.

First of all, the repeated claims for the special status of chymical arcana are *topoi*—literary conventions used as matters of course by virtually all early modern alchemical authors. Chapter 2 recounted how such
conventions emerged with the initiatic style of the Jābirian corpus in
the ninth and tenth centuries, and were perpetuated by medieval Latin
authors who mimicked Arabic styles to give their writings an air of
venerable authority. Such mimicry included not only the initiatic style
stemming from the Isma'ili sources but also the inclusion of Latin equivalents of the ubiquitous Arabic phrase *insha'allah* (if God wills). Hence,
curiously enough, the religious tone of some Christian texts was actually determined in part by Muslim expressions of piety. Later European
authors, who were often members of religious orders or at least pious

laymen, further developed and extended these topoi, which had by then become virtually automatic.

Second, the term *donum dei* (gift of God) is actually a technical phrase used in medieval and Renaissance theological and legal literature deal-ing with the status of knowledge. St. Thomas Aquinas (among others) asserts that *all* knowledge is in fact a *donum dei*. In so doing, he alludes to an established legal precept that "knowledge is the gift of God, therefore it cannot be sold" (scientia donum dei est, unde vendi non potest).[43] This precept emerged from ethical arguments about whether it is licit for teachers to require payment from their students. (The consensus was *no*.) The underlying idea is that since knowledge is a divine gift, the per-son who has received it has no right to sell it, in part because he does not in fact *own* it, and in part because doing so would be simony, that is, the sin of selling spiritual goods for money. Late medieval and early mod-ern chymical writers were surely aware of the term's background. Their use of it both underscores the ultimate source of all knowledge while of course elevating theirs in particular, and emphasizes the obligation to use the gift of knowledge wisely and appropriately.[44]

Third, and most significantly, moderns tend to make disconnections where early moderns did not—in this case, between science and religion. Moderns like to keep the two conveniently boxed up separately, prefer-ably at a safe distance from each other. Convinced that such modern conventions are somehow intrinsically normal, many readers nowadays tend to think that chrysopoetic authors who speak of their subject as a holy gift of God are somehow *ab*normal, that is, in need of special expla-nation. But perhaps moderns are the ones who require special explana-tion: where, why, and by whom did the presently accepted boundaries of disciplinary identities come into being? Early moderns, who recog-nized God as the Creator of all things, recognized all things as His gifts. Authors, both chymical and otherwise, knew and quoted the biblical passage that "Every worthwhile gift, every genuine benefit comes from above, descending from the Father of Lights" (James 1:17). Moderns, on the other hand, tend to imagine divine activity or presence as an ex-ceptional event, something disconnected from daily life by a respect-able amount of thunder and lightning, like something out of a Cecil B. DeMille extravaganza. But for the early modern, the action and presence of the divine was something constant, daily, even familiar.

It therefore disrupts comfortable modern categories to thumb through one of George Starkey's laboratory notebooks and find, amid careful records of the exact dates of his experiments, the weights of materials he used, and the length of time he heated them, an entry like this:

> At Bristol, on 20 March 1656, God revealed to me the whole secret of the liquor alkahest; let eternal blessing, honor, and glory be to Him.[45]

Here Starkey very matter-of-factly records—as if he were describing some important but not terribly surprising experimental outcome—his reception of a *donum dei*. While this terse entry does not specify exactly *how* this gift came to him, in another notebook he describes a sequence of logically connected experiments that he calls the result of a "divine nod" (*divino nutu*). In yet another, he recounts, "I completed many unfruitful trials, yet finally God deigned to direct me into the true art."[46] Starkey is clear that the knowledge acquired in his laboratory operations was a divine gift, but it was so in the sense that God was ever present, nudging and hinting providentially towards discoveries in subtle ways behind the ordinary appearances of things. Here there is neither a God speaking melodramatically from the clouds in stentorian tones nor a chymist swept up into an ecstatic state. Starkey's notebooks depict instead a chymist giving thanks to God for that mysterious moment of intellectual clarity we sometimes call the "eureka moment." The chymist labors diligently, remaining always conscious of divine omnipresence and providence, and acknowledges his Creator as the ultimate source of knowledge. We might judge this perspective as making divinity routine, debasing it into a part of the workaday world; but an equally plausible view, and one more in tune with early modern conceptions, is that it renders the world and human striving divine by raising them into constant but silent connection with the transcendent.

Such a connection was one thread in a web of connections envisioned to link human beings, the natural world, and the divine at multiple levels.[47] This early modern vision of an interlinked cosmos is depicted graphically in an intricate image designed by Robert Fludd (1574–1637). Fludd was an English physician, scholar, and philosopher who engaged in debates with some of the most prominent thinkers of his day. He also wrote about topics in chymistry, including the Philosophers' Stone.[48]

The elegant engraving shown in figure 7.6*a* was produced in 1617, probably by Matthaeus Merian, the same artist Michael Maier employed.

At the center of the image is the Earth; and seated atop it is an ape (fig. 7.6*b*). This ape represents human artifice, often called the "ape of nature," because it imitates ("apes") nature's work. The ape's height defines four concentric circles representing the varieties of human knowledge. Moving outward from the Earth, the first circle describes "art correcting nature in the mineral realm," with tiny depictions of distillations to represent chymistry—the artifice that corrects the imperfections of material substances by transmuting base metals into gold, by turning toxic materials into medicines, or by simply transforming the common into the useful and unusual. The next circle shows "art aiding nature in the vegetable realm," exemplified by agriculture and the grafting of fruit trees. Thereafter follows "art supplementing nature in the animal realm," instanced by such things as medicine, raising silkworms, and incubating eggs artificially, and (according to an ancient belief) spontaneously generating bees from a bull carcass to start new hives. These three types of practical arts all share the theme of improving on nature so evident throughout alchemy. In the fourth circle stand the "more liberal" arts—meaning those that are freer from servitude to utilitarian production. In Fludd's rather idiosyncratic list, all of them are mathematically based: astronomy, music, geometry, timekeeping, painting, fortification, and so forth.

The realm of human activity represented by the ape and the four circles is inescapably tied to the rest of the universe. The ape is chained by its wrist to a female figure standing above it representing nature; nature controls what human artifice can do. The concentric realms that she defines contain (reading outward) minerals and metals, plants, and animals including man. Rising farther, above the terrestrial, come the celestial bodies, the concentric circles of the seven planets that revolve around the Earth in the geocentric view of the cosmos. Here Fludd draws in a few lines of correspondence as examples. On the left side, the planet Saturn is connected by two lines to lead and antimony in the sphere of minerals. Similarly on the right, Venus is connected to copper and orpiment. Again on the left, the Sun is connected to man, whose arms open to the connection (both are hot and dry). Symmetrically on the right, woman receives the influence of the analogously cold and wet Moon, whose monthly cycle she imitates in her own body.

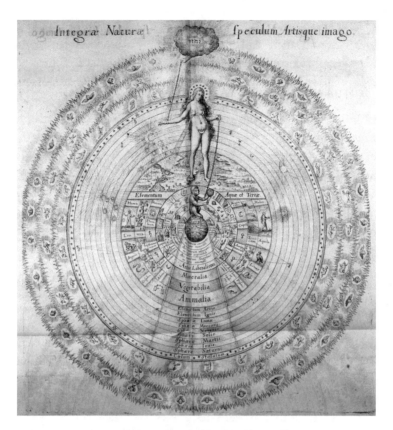

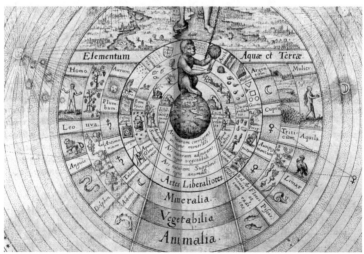

Figure 7.6a, b. *The Mirror of the Whole of Nature and the Image of Art*, from Robert Fludd, *Utriusque cosmi historia* (Oppenheim, 1617). Courtesy of the Roy G. Neville Historical Chemical Library, Chemical Heritage Foundation, Philadelphia.

Equally significantly, nature herself is no more independent than human artifice, for her own wrist is connected to a chain held by a hand descending from a numinous cloud positioned above the ranks of angels beyond the physical universe, which is bounded by a circle of stars. The cloud bears the Sacred Tetragrammaton, the four Hebrew letters expressing the unutterable Name of God. Thus, the emblem portrays every act of human industry as chained to nature, and nature in turn chained to the hand of God. The entire cosmic system is linked into a complex, interconnected, interfunctioning whole. In this view, the chymist's work in his laboratory—depicted in the lowest circle—remains always connected to divine will and providence, that is to say, reliant on divine gifts and guidance, as is the work of the farmer, the physician, and the astronomer.

This vision of a tightly interconnected cosmos stems from multiple sources deeply embedded in Western culture. Neoplatonic thought stressed the notion of a "ladder of nature," where everything in existence, from inanimate matter to the transcendent One, is linked in a hierarchical chain. Aristotle, in his natural philosophy, endeavored to be comprehensive; his ideas about motion, causation, the qualities, and so forth applied uniformly throughout the various realms of nature. The long-standing idea of celestial effects on earth—the macrocosm-microcosm interaction expressed in the *Emerald Tablet* ("as above, so below") and that undergirded astrology—was regarded as visible every day in the tides, the seasons, and the turning of the compass toward the North Star; such interactions joined things celestial and terrestrial. Perhaps most powerfully, the Christian faith in a single, omnipotent, providential, and omniscient God implied that the world is a uniform, harmonious, interfunctioning whole—a *cosmos* in the true sense of the word, that is, a highly ordered whole—for it is the product of a single, perfectly self-consistent Mind. The creation reflects the unity of its Creator.

With this larger context in place, we can better understand the remarkable Heinrich Khunrath (circa 1560–1605).[49] Khunrath's wide-ranging interests and activities further illustrate the connections within early modern thought, particularly in the area of chymistry and religion. He was certainly involved in practical chrysopoeia, and he also affirmed the value of theurgical practices—that is, ritual methods for calling down divine revelations in dreams and visions—for gaining knowledge not only about chrysopoeia but about other topics as well. Not surprisingly, therefore, he explicitly states that knowledge of the Philosophers' Stone is a *donum dei*. But what does that term really *mean* for him? Khunrath asserts that the gift consists of two main secrets—the same two described in chapter 5: knowledge of the correct starting material and knowledge of how to treat it practically to produce the stone. After reiterating that the knowledge of such secrets is truly a gift of God (*Gabe und Geschenck Gottes*), Khunrath continues:

> The dear ancient philosophers obtained the knowledge and practice of this thing [the Stone and its matter], as one can clearly find in their books, either from God himself through a special divine inspiration, secret visions, or the revelations of good spirits, or from another philosopher and human teacher, or indeed from the light of nature, through diligent reading, true books, and by contemplating, meditating upon and observing wisely the wonderful workings of nature in the larger world.[50]

Read that list again slowly. Khunrath places side by side things that moderns would separate radically. Divine revelation and visions are laid alongside ordinary human instruction, the study of books, and careful observations of the world at large. The connection among them is that God is ultimately the source of knowledge. All knowledge comes from Him, whether immediately; mediately but supernaturally through an angelic vision; or mediately and naturally through the voice of a human teacher, the words of a book, or the sight and investigation of creation itself. While we moderns would separate divine activity into a special category, for Khunrath (or Starkey, or Fludd, or so many of their contemporaries) it was simply another way of acquiring knowledge—or to say it better, divine activity was seen ultimately as the foundation for *all* methods of acquiring knowledge. "Every worthwhile gift, every genuine benefit comes from above, descending from the Father of Lights." Such an outlook, in short, was a result of the early moderns' sense, their awareness, of the constant presence of the divine in their daily lives and in their world. Here is part of the world we have lost. Until we regain it, we cannot really understand the way early modern people thought and lived.

Traces of this worldview survive in modern colloquialisms. When an idea or answer to a problem suddenly comes into our heads, we are apt to call it a "moment of inspiration," even if our intention is divorced from the theology implicit in that word. Where *do* creative ideas really come from, anyway? The early moderns would say they ultimately come from the Great Source of all creation. Thus, the acquisition of rare knowledge—howsoever obtained—is truly a gift of God. But that gift need not be wrapped in thunder and lightning or in ecstatic visions (although it might be); it can arrive silently and gently while reading a book, listening attentively to a teacher, contemplating the actions of nature, or bending over a crucible.

The connections existing within the complex of nature, God, and human beings can serve a further and perhaps more surprising function. Khunrath's treatment of the Philosophers' Stone begins, as is typical, with proofs that it actually exists. He appeals first to testimony, what he calls the "repeated experience" offered by past possessors of the stone. He then cites the support of chymical theory, what he calls the "consensus of rightly philosophizing chymists." But he then adds a third proof, one that he calls the most convincing, namely,

> the wonderful harmony of the Philosophers' Stone with Jesus Christ, and vice versa, placed verily by God not vainly before our eyes. If a Christian heart should rightly consider nothing more than this single witness by itself, or be instructed regarding it, it would thereupon necessarily prefigure and testify regarding the possibility of the natural Philosophers' Stone, that the blessed and heavenly Stone exists in nature, from the beginning of the world.[51]

Here Khunrath claims that Christ guarantees the reality of stone. What on earth is he talking about? How does he justify this leap? In part, Khunrath is drawing on the familiar allegory linking Christ and the Philosophers' Stone—a connection developed in the late Middle Ages, in the writings of pseudo-Arnald of Villanova, John of Rupescissa, and the *Rosarium philosophorum*. In making the stone, the prepared ingredients are heated in the *ovum philosophorum* until they blacken in "death"; further heating "revives" them into a new, glorified, and subtilized substance in a process akin to "resurrection." This finished stone can then "heal" the defects and imperfections of the base metals, like the risen Christ redeems the world by healing a fallen and imperfect humanity and creation.

For Khunrath, however, this comparison—or "analogical harmony" (*harmonia analogica*), as he terms it—between Christ and the Philosophers' Stone is much more than a metaphor, allegory, or rhetorical conceit. It carries *demonstrative, evidentiary, and probatory power.* The existence of Christ the Redeemer and His attributes guarantees the existence of a material stone with analogous attributes relative to its own material realm. The connective analogy (Christ-Philosophers' Stone) functions as a *proof*, transmitting the sure existence of one to the sure existence of the other. How can this be? On the one hand, it is an ultimate expression of the axiom "as above, so below." On the other, it expresses a keen difference between modern and early modern understandings of metaphors and analogies. The modern world considers such metaphors and analogies to be creations of the human mind. For Khunrath and many of his contemporaries, they are neither arbitrary nor products of human imagination—they exist independently as real connections in the fabric of the world itself. They lie there hidden, waiting to be uncovered.

Almost three centuries earlier, Petrus Bonus said nearly the same thing, although he followed the connection from the Philosophers' Stone to Christ in the other direction. Bonus asserted that pre-Christian adepts used their observations in making the stone to foretell the Messiah's virgin birth.[52] Closer to Khunrath's own time, Robert Boyle likewise used laboratory observations of the recovery of original starting materials at the end of a sequence of chymical operations as evidence for the Christian doctrine of the resurrection of the body.[53] Other authors did the same, based on palingenesis. The physician Sir Thomas Browne (1605–1682) writes that what he learned about the Philosophers' Stone "hath taught me a great deal of divinity, and instructed my beliefe, how that immortall spirit and incorruptible substance of my soule may lye obscure, and sleepe a while within this house of flesh."[54]

Pierre-Jean Fabre (1588–1658), a physician, chymist, and prolific author on all aspects of chymistry including chrysopoeia, produced the remarkable *Christian Alchemist* (*Alchymista christianus*) in 1632, perhaps the most extended use of chymical observations to point at theological truths. The book's purpose is to explicate "as many mysteries as possible of the Christian faith through chemical analogies and figures," and to demonstrate "the orthodox doctrine, life, and virtue of Christians using the chymical art.[55] While Fabre's aims here seem superficially akin to the

interpretation of alchemy proposed by Ethan Allen Hitchcock in the
nineteenth century, there is a profound difference. Fabre did not reduce
alchemy entirely to theological allegory; rather, he saw practical laboratory work and phenomena as coexistent and coextensive with theological
truths, and naturally linked to them. Chymistry expresses and corroborates theological truths. Would not the Creator have implanted analogical or allegorical images of Himself throughout His creation, where they
could be discovered by human beings? This vision of the world rests
in part on the doctrine of the Two Books, enunciated most fully by St.
Augustine (354–430), and widely accepted in the early modern period
by theologians and natural philosophers alike. The doctrine states that
God reveals Himself to mankind in two different ways: through words
in the the Bible, the Book of Scripture, and through things in creation,
the Book of Nature. Hence, investigation of the natural world, through
chymistry, for example, automatically entails uncovering "a greate deal
of divinity."

Crucially, these ideas and visions are not idiosyncratically alchemical.
Similar views and arguments exist throughout early modern thought.
For example:

> In the sphere, which is the image of God the Creator . . . there are three
> regions, symbols of the Three Persons of the Holy Trinity: the center, a
> symbol of the Father, the surface, of the Son, and the intermediate space,
> of the Holy Spirit. So, too, just as many principal parts of the world have
> been made: the sun in the center, the sphere of the fixed stars on the sur-
> face, and the planetary system in the intermediate region.[56]

This is no "alchemist" speaking, but rather the famous astronomer
Johannes Kepler (1571–1630), renowned for his laws of planetary motion
that remain standard fare in physics and astronomy courses today. The
passage cited here is part of his argument in favor of Copernicus's helio-
centrism, the idea that the Sun is at the center of the cosmos, with the
Earth in motion around it, not the other way around. Kepler chooses
to argue not from observational evidence but rather by using *harmonia analogica*, to borrow Khunrath's phrase, an analogical connection
between the physical universe and its invisible Creator. The attributes
of God guarantee a heliocentric universe. God the Father is the source
and immutable origin from all eternity, and thus His physical symbol,

His analogue or metaphor in the world, the Sun, *must* rest at the center, illuminating, warming, and invisibly guiding all the planets, including the Earth. The analogical harmony is evidence of the way things are. Throughout his works, in fact, Kepler "traverses the labyrinths of the mysteries of nature by using the thread of analogy."[57]

The connectedness of all knowledge, the interconnections of nature, God, and human beings, and the evidentiary power of analogy appear also in the work of Jesuit polymath Athanasius Kircher (1601 or 1602–1680). The emblematic frontispiece (fig. 7.7) to his 1641 book *The Magnet* displays these connections and summarizes them with the statement "All things rest connected by hidden knots" (Omnia nodis arcanis connexa quiescunt). Emblems bearing the names of various kinds of knowledge (astronomy, philosophy, perspective optics, music, theology, medicine, and so on) are shown arranged in a circle, all linked by chains. They in turn are chained to three larger circles within the one they form: the sidereal world (everything farther away than the Moon), the sublunary world (the Earth), and the microcosm (that is, human beings). These three divisions of the universe are themselves chained together, and at their center, touching each one of them equally, stands the *mundus arche-typus*—the archetypal world, that is, the mind of God that contains the models of everything possible in the universe.

For Kircher, these invisible connections or "hidden knots" are exemplified in the magnet's invisible power over iron.[58] He begins his book, then, with an exhaustive description of magnets and their effects, and then expands outward to other objects that display similar "magnetic" effects: from static attraction, to the turning of a sunflower toward the Sun throughout the day, to sympathetic vibrations, to the sympathy and antipathy between certain plants and animals, to the motions of planets in their orbits, and so on. Slowly Kircher ascends from example to example—in a manner often bizarre to us, and even to some of his contemporaneous readers—until finally he emerges beyond the limits of the physical universe, and he connects all these phenomena to God's invisible but inescapable love, the one true original force that binds everything that is, attracting "magnetically" the whole of creation back to its Source. According to Kircher's teaching, then, we can *witness* divine love every day in the action of the magnet that clips notes to the refrigerator.

For the early modern thinker, such analogies—or metaphors, or harmonies, call them what you will—meant vastly more than they do to

Figure 7.7. Frontispiece to Athanasius Kircher, SJ, *Magnes sive de magnetica arte* (Rome, 1641), showing the interconnectedness of all knowledge, the natural world, human beings, and God. Courtesy of the Division of Rare and Manuscript Collections, Cornell University Library.

moderns. For them, an analogy was something actually existing in the world—a real connection intentionally built into the fabric of what is. Metaphors and analogies constituted a central facet of their multi-layered, multivalent, highly interconnected world. The power of such *harmoniae analogicae* flowed from their vision of a world created by a uniform, omnipotent, omniscient God; a world endowed in every corner with meaning, message, and purpose; a world where heaven is joined to earth, and God to man (His *image*) in ways seen and unseen, ways to be discovered and explored by manifold means. Thus, an analogical likeness was not the product of a poetical human mind but a line in the blueprint of creation.

With this vision in mind, Khunrath's proof for the existence of the Philosophers' Stone becomes clearer, and points toward a deeper understanding of the early modern world more generally. The complex of connections and symmetries existing in the world undergirded a layered, multivalent meaning for everything in it. Contemporaneous artistic productions—paintings, literary works, and music—based themselves on a love of layered meaning and allegory, meanings not found on the surface but teased out by the viewer. Educated early moderns *expected* multiple levels of meaning in their art, literature, and theatre, and they delighted in seeking and finding them. Crucially, natural philosophers of the period generally expected the very same multivalent meaning to exist not only in human creations but to an even *greater* extent in God's creation—the natural world itself. As the ultimate Author and Artist, God created the world at every level as the ultimate baroque masterpiece of layered, multivalent, allegorical, and symbolic meaning. Hence, observations of the natural world carried meanings far beyond the isolated literal object under immediate consideration.

In terms of alchemy specifically, what I am arguing is that the connections between it and the divine are indeed very close, but that relationship was not wholly unique in the early modern period. A similar close relationship can be found in other studies of the natural world from that time (as in Kepler and Kircher). These relationships are manifestations not only of the piety but more intrinsically of the unified cosmic vision of so many thinkers of the period. The significant links between alchemy and religion might sometimes seem to mark it out as something alien from "science," but only if the comparison is made with reference to the science of our own day. When recognized figures like Kepler, Boyle, or Newton are rightly contextualized, they too cease to fit comfortably into modern conceptions of science. They do, however, fit comfortably—alongside alchemy—within the *natural philosophy* of their day, that comprehensive study of the world in its interconnected entirety, embracing human beings, nature, and God.[59] The goal of much early modern thought involved finding, building, and using connections in the world—not, as is so often the case in modern science, dissecting and isolating things to be studied solely in and of themselves. Seen from the early modern perspective of a unified cosmos, the grand unification theory sought by modern physicists (the way chymists sought the Philosophers' Stone) is a nice idea but ultimately a blinkered trifle, because it embraces so little and ignores so much. I suggest that alchemy can seem strange to us in part because it reflects the larger context of early modern thought as it existed before the narrowing of natural philosophy into science, and employs ways of thinking and seeing that have not been transmitted to us within the methodologies and metaphysics of modern science.

Having sampled the dynamism and diversity of chymistry, and its connections to so many other fields of knowledge and creativity, it is hardly surprising that chymistry enjoyed so wide a dispersion in early modern culture. Throughout various branches of human endeavor, it fired the imagination of artists, authors, theologians, and natural philosophers alike, because it shared so many visions and aims with them. Early modern chymistry with its arresting images and ideas (once we understand them properly and contextually) can tell us much about the general outlooks of the premodern world, a world of which and from which we still have so much more yet to learn.

❁ EPILOGUE ❁

The idea of turning the common into the precious captures the imagination; chrysopoeia and other alchemical endeavors embody this fascination. Yet alchemy is more than gold making, more even than the transformation of one substance into another. From the time of its emergence in Greco-Roman Egypt nearly two thousand years ago and down to the present day, it has evolved in a variety of cultural and intellectual contexts, and developed along multiple lines. A myriad of practitioners pursued it for various reasons along many pathways toward a variety of goals. The range of ideas and practices outlined in the preceding chapters complicates the problem of answering the fundamental question of what alchemy was really about. No simple response fully suffices. But recognizing such diversity and dynamism—both over time and at any given time—reveals alchemy's identity in more interesting and more historically accurate ways. Amid such an array of practices, goals, ideas, and practitioners, however, a few relatively stable features of the Noble Art do emerge.

First and foremost, alchemy was an endeavor of both head and hand. It was both theoretical and practical, textual and experimental, and these two aspects constantly interacted. Theories about matter and its composition—Zosimos's "soul and body," Jābir's Mercury and Sulfur, Geber's *minima*, Paracelsus's *tria prima*, the Scholastics' prime matter and substantial form, Van Helmont's *semina*, and all the rest—undergirded alchemical aims and directed practical laboratory endeavors.

Observations in the laboratory and in the wider world formed a core of experiences from which such theories sprang and continued to develop. The existence of these theories and their role in practical work discredit the old notion that alchemy was no more than trial-and-error cookery.

Conversely, alchemical laboratory practices and results—described both clearly and not-so-clearly in text after text, concealed and revealed in allegory and emblems, and witnessed by surviving artifacts—equally discredit the notion that alchemists inhabited a merely speculative world, or that their immediate aims were not material ones. Alchemists pored over the writings of their predecessors for the purpose of putting them into practice, constantly reinterpreting and adding to them from their own experience. The broad spectrum of alchemists certainly ranged from armchair theorizers at one end to narrow recipe-followers at the other, but alchemy's core depended on the interactions of theory and practice. Straddling the otherwise disparate realms of the artisan and the intellectual, it thrived as an investigative enterprise for exploring the world and its possibilities; its goals included both knowing and doing.

With its emphasis on practical work, alchemy was also a *productive* enterprise. Producing new materials and transforming or improving common ones forms a central theme within the alchemical tradition. The products alchemists sought to prepare ranged from grand arcana like the Philosophers' Stone, the alkahest, and potable gold through lesser transmuting agents, spagyric and other pharmaceutical preparations, to greater yields of metal from ores, better alloys, pigments, glass, dyes, cosmetics, and a host of other commercial products. Some practitioners focused their efforts on preparing just one or two of these products, while others turned their attention and expertise to more or even all of them. This emphasis on producing materials often earned alchemy the scorn of more bookish observers, but it resulted in a special degree of physicality unmatched by any other subject outside the artisanal trades. It also resulted in the development and accumulation of methods for manipulating, identifying, and analyzing substances—comprising a rich store of "how-to" knowledge.

Alchemical productivity was not limited to physical products; it also aimed to produce knowledge about the natural world. Working with and transforming matter required knowing about what it really was, theorizing about its hidden nature and composition, and understanding its properties. Alchemists' experiences led them to formulate, for example, hypotheses about unseen, semipermanent microparticles of matter that lay at the heart of material transformations and that could explain their observations. They noted the conservation of weight of the materials used in their experiments and relied on it to monitor their results better than unaided senses alone could do. They cataloged substances and their properties, recording the fullness and diversity of the natural world. In short, they sought to *understand* the natural world, to uncover, observe, and utilize its processes, to formulate and refine explanations of its functioning, and to seek out its arcane secrets.

Crucially, the "natural" world was not so neatly circumscribed for early modern people as it is for moderns. In a world filled with meaning, where human beings, God, and nature are profoundly intertwined on multiple levels, the alchemists' laboratory investigations and findings had wider scope and ramifications than do the analogous activities of today's chemists. Within this wider scope, theological and natural truths could reflect and expound on one another, and the study of nature was the study of God at one remove. Hence, alchemy possessed a multivalency that operated across multiple branches of knowledge and culture. Small wonder, then, that it inspired not only other investigators of nature but also a range of artists and authors (even to the present day) who could find meanings of their own in its claims, promises, and language. Thus, alchemy forms a part of not only the history of science, medicine, and technology but also the history of art, literature, theology, philosophy, religion, and more. These diverse cultural connections and its multivalent character distinguish alchemy—as well as contemporaneous astronomy, natural history, and other natural philosophical pursuits—from more narrowly focused modern sciences.

Yet alchemy, as an integral part of natural philosophy, remains foremost a part of the long history of science, of the endeavor of human beings to know, understand, control, and make use of the world. Its difficult textual legacy as well as long-lived misconceptions or misrepresentations of its aims and practitioners often obscured this connection, but current scholarship restores the continuities (without ignoring the important distinctions) between alchemy and modern science. The alchemists' insistence on practical work linked with theoretical speculation promoted a culture of experimentalism and developed investigative methodologies (such as analysis and synthesis) crucial to the modern scientific enterprise. Alchemists' aspirations to produce gold and silver, gems, better medicines, and other products argued for the power of human artifice to improve on nature. Consequently, no clean "rupture" separates alchemy from chemistry. To be sure, goals, theories, and worldviews as well as social and professional structures and cultural positions changed, usually gradually, but the focus on understanding matter and guiding its transformations toward practical ends establishes a commonality and continuity between "alchemy" and "chemistry." We might ponder whether today's chemist is any more distant from, say, George Starkey, than Starkey was from Jābir, or Jābir from Zosimos. Although these individuals would undoubtedly be confused (and often confounded) by each other's specific ideas and theories—not to mention cultural assumptions—I think they would probably recognize amid such differences a certain kinship linking them into a long "chymical" tradition of questions about and desires to manipulate the material world. Of course, many ideas developed and held by the practitioners of *chemeia*, *al-kīmiyā'*, *alchemia*, *chymistry*, and chemistry have subsequently been shown to be factually incorrect. Nonetheless, science is not a body of facts existing "out there"; it is an ever-developing story about the world as told by human observers rooted in time and place. Chymists were (and continue to be) important authors of that story.

When the pioneering historian of alchemy Frank Sherwood Taylor wrote his popular 1952 survey *The Alchemists*, he modestly referred to it as only an "interim report" based on what he saw as a still very incomplete state of knowledge about the subject. Now, sixty years later, we are in possession of a far broader and deeper understanding of it. An energetic battalion of scholars of alchemy has been busily extending our knowledge, and alchemy has been brought back into the fold of serious academic study and discourse. Yet, as I write these final lines, I recall the tens of thousands of pages of alchemical books and manuscripts I've encountered in countless libraries and archives, so many of which have not been carefully read in centuries. Even a glance up at my own bookshelves reveals rather forbiddingly large volumes, much of whose tiny early-modern typography still awaits being brought to life and into our story by knowledgeable eyes and hands. *Ars longa, vita brevis.* There is no danger that all the secrets of alchemy have been revealed here, or in any other book. We still have much to learn, and the Noble Art still has much to teach.

❦ ACKNOWLEDGMENTS ❦

The idea of writing *The Secrets of Alchemy* originally emerged from conversations during a series of three lectures I gave at Oregon State University when I had the honor and pleasure of being invited there as the 2008 Horning Visiting Scholar. Paul Farber and Mary Jo Nye read the first drafts of this manuscript, and their comments and encouragements were crucial to its development and to the continuation of this project. My sincere thanks go first and foremost to them and to the Horning Foundation.

Throughout this book's overly long gestation, many other friends and colleagues read all or part of the numerous intermediate drafts and shared generously of their expertise and knowledge: Dane Daniel, Peter Forshaw, Renko Geffarth, Benjamin Hallum, Wouter J. Hanegraaff, Didier Kahn, K. D. Kuntz, Marcos Martinón-Torres, Matteo Martelli, Bruce Moran, William R. Newman, Margaret J. Osler, Albert Philipse, Jennifer Rampling, Sean T. Schifano, James Voelkel, and the students in my graduate seminar "Wretched Subjects," taught at Johns Hopkins in spring 2012. I am extremely grateful to them for their critiques, corrections, help, and patient support.

I would also like to express my thanks to the generous people at several institutions who provided images, often of unique objects or texts: James

Voelkel, Amanda Shields, and Marjorie Gapp at the Chemical Heritage Foundation; Susan Stravinski and Jill Rosenshield at the University of Wisconsin–Madison Libraries; Earle Havens at the Sheridan Libraries of Johns Hopkins University; and David Corson at Cornell University Library.

❋ NOTES ❋

Introduction

1. Some general historical surveys of alchemy in English include John Read, *Prelude to Chemistry: An Outline of Alchemy, Its Literature and Relationships* (originally published 1936), E. J. Holmyard, *Alchemy* (originally published 1957), and Frank Sherwood Taylor, *The Alchemists: Founders of Modern Chemistry* (originally published 1949); the last is the best of them. These were useful introductory texts in their day, but much of their contents have been superseded by more recent scholarship.

2. See Nathan Sivin, *Chinese Alchemy: Preliminary Studies* (Cambridge, MA: Harvard University Press, 1968), which contains useful essays as well as edited texts, translations, and some chemical explanations, including several replications of processes, and "Research on the History of Chinese Alchemy," in *Alchemy Revisited*, ed. Z. R. W. M. von Martels (Leiden: Brill, 1990), pp. 3–20. Also, Joseph Needham, *Science and Civilisation in China*, vol. 5, *Chemistry and Chemical Technology*, esp. parts 2–5 (Cambridge: Cambridge University Press, 1974–83), and Hong Ge, *Alchemy, Medicine, Religion in the China of AD 320* (Cambridge, MA: MIT Press, 1967). Needham's groundbreaking work needs to be read with circumspection, since he is sometimes prone to making arbitrary definitions as well as broad claims about Chinese influence on the West, as in "The Elixir Concept and Chemical Medicine in East and West," *Organon* 11 (1975): 167–92. Indian alchemy remains less studied; see Praphulla Chandra Ray, *A History of Hindu Chemistry*, 2 vols. (London: Williams and Norgate, 1907–9; reissued and expanded as *History of Chemistry in Ancient and Medieval India* [Calcutta: Indian Chemical Society, 1956]), and Dominik Wujastyk, "An Alchemical Ghost: The Rasaratnakara by Nagarjuna," *Ambix* 31 (1984): 70–84. Both topics are in need of careful study and reevaluation.

Chapter One

All translations in the text from Latin, German, French, Italian, and Greek are my own unless otherwise noted.

1. Alfred Luca and John R. Harris, *Ancient Egyptian Materials and Industries* (London: Arnold, 1962); Martin Levey, *Chemistry and Chemical Technologies in Ancient Mesopotamia* (Amsterdam: Elsevir, 1959); Marco Beretta, *The Alchemy of Glass: Counterfeit, Imitation, and Transmutation in Ancient Glassmaking* (Sagamore Beach, MA: Science History Publications, 2009), 1–22; Peter van Minnen, "Urban Craftsmen in Roman Egypt," *Münstersche Beiträge zur antiken Handelsgeschichte* 6 (1987): 31–87; Paul T. Nicholson and Ian Shaw, eds., *Ancient Egyptian Materials and Technology* (Cambridge: Cambridge University Press, 2000); Fabienne Burkhalter, "La production des objets en métal (or, argent, bronze) en Égypte Hellénistique et Romaine à travers les sources papyrologiques," in *Commerce et artisanat dans l'Alexandrie hellénistique et romaine*, ed. Jean-Yves Empereur (Athens: EFA, 1998), pp. 125–33; and Robert Halleux, *Le problème des métaux dans la science antique* (Paris: Les Belles Lettres, 1974).

2. The most recent and reliable edition of these papyri (with French translation) is Robert Halleux, *Les alchimistes grecs I: Papyrus de Leyde, Papyrus de Stockholm, Recettes* (Paris: Les Belles Lettres, 1981). There exist older English translations in Earle Radcliffe Caley, "The Leiden Papyrus X: An English Translation with Brief Notes," *Journal of Chemical Education* 3 (1926): 1149–66, and "The Stockholm Papyrus: An English Translation with Brief Notes," *Journal of Chemical Education* 4 (1927): 979–1002.

3. Halleux, *Les alchimistes grecs*, pp. 104–5. The Greek name of the substance is ambiguous; in many contexts it can be translated as either "water of sulfur" or "divine water"; see below.

4. Should any readers wish to try this for themselves, take calcium hydroxide (5 g) and sulfur (5 g), and mix with 100 ml of fresh urine (if you're squeamish about that, try 100 ml of distilled white vinegar instead). Boil very gently in a well-ventilated space for one hour, and filter the solution while hot. It takes a bit of trial and error to use the liquid effectively, but the surface color produced can be surprisingly stable and long lasting.

5. This body of Greek texts was edited (with French translation) by the chemist Marcellin Berthelot and C. E. Ruelle in their *Collections des alchimistes grecs*, 3 vols. (Paris, 1887–88). Their pioneering work has often been criticized, and on reasonable grounds; the translations are frequently unsound, and the Greek texts often inaccurate. Yet it remains the only available source for many of the texts, since only some of them have received better attention since. On the manuscripts, see Michèle Mertens, *Les alchimistes grecs IV, i: Zosime de Panopolis, Mémoires authentiques* (Paris: Les Belles Lettres, 2002), pp. xx–xlii; Henri Dominique Saffrey, "Historique et description du manuscrit alchimique de Venise *Marcianus graecus* 299," in *Alchemie: Art, histoire, et mythes*, ed. Didier Kahn and Sylvain Matton, Textes et Travaux de Chrysopoeia 1 (Paris: SÉHA; Milan: Archè, 1995), pp. 1–10; and A. J. Festugière, "Alchymia," in *Hermétisme et mystique païenne*, ed. A. J. Festugière (Paris: Les Belles Lettres, 1967), pp. 205–29. For an extensive list of Greek alchemical manuscripts, see Joseph Bidez et al., eds., *Catalogue des manuscrits alchimiques grecs*, 8 vols. (Brussels: Lamertin, 1924–32).

6. Matteo Martelli, "L'opera alchemica dello Pseudo-Democrito: Un riesame del testo," *Eikasmos* 14 (2003): 161–84; "Chymica Graeco-Syriaca: Osservationi sugli scritti

alchemici pseudo-Democritei nelle tradizioni greca e sirica," in *'Uyūn al-Akhbār: Studi sul mondo Islamico; Incontro con l'altro e incroci di culture*, ed. D. Cevenini and S. D'Onofrio (Bologna: Il Ponte, 2008), pp. 219–49; and Christoph Lüthy, "The Fourfold Democritus on the Stage of Early Modern Europe," *Isis* 91 (2000): 442–79. An English translation of the *Physika kai mystika* was published in 1890, but it is incomplete and often misleading: Robert B. Steele, "The Treatise of Democritus on Things Natural and Mystical," *Chemical News* 61 (1890): 88–125. A much-needed critical edition with Italian translation has recently been published by Matteo Martelli, ed., *Pseudo-Democrito: Scritti alchemici, con il commentario di Sinesio; Edizione critica del testo greco, traduzione e commento*, Textes et Travaux de Chrysopoeia 12 (Paris: SÉHA; Milan: Archè, 2011); an English version including new material from Syriac versions is now in preparation by the same scholar. Martelli (pp. 99–114) also disposes of the earlier notion, frequently repeated in the literature, that the *Physika kai mystika* was written by one Bolos of Mende, a Greco-Egyptian author of the third and second centuries BC.

7. Originally, the word was used in relation to the material details of religious rituals, but by the start of the Christian Era it had come to refer to anything that required laborious activity to uncover. Louis Bouyer, "Mysticism: An Essay on the History of a Word," in *Understanding Mysticism* (Garden City, NY: Image Books, 1980), pp. 42–55.

8. Martelli, *Scritti alchemici*, pp. 184–87.

9. See the linguistic analysis by Matteo Martelli in "Greek Alchemists at Work: 'Alchemical Laboratory' in the Greco-Roman Egypt," *Nuncius* 26 (2011): 271–311, esp. 282–84.

10. The most reliable and extensive treatment of Zosimos's Greek texts is Mertens, *Les Alchimistes grecs IV, i: Zosime*. Some Zosimos texts not treated by Mertens but published by Berthelot, *Collections*, 117–242, await more critical editions.

11. Recent scholarship suggests that Zosimos organized his writings toward the end of his life, classifying his writings under the twenty-four letters of the Greek alphabet and adding prologues to each (either as introductions or as responses to criticism). He then added four final books to create the ensemble of twenty-eight alluded to by the *Suda*, a tenth-century Byzantine encyclopedia. We currently possess fragments known to have been classed under omega, and references to sigma and kappa. See Mertens, *Les Alchimistes grecs IV, i: Zosime*, pp. ci–cv.

12. A careful and insightful analysis of Zosimos's apparatus, including clear illustrations, is given in ibid., pp. cxiii–clxix; see also Martelli, "Greek Alchemists."

13. Mertens, *Les Alchimistes grecs IV, i: Zosime*, p. 12. The whitening by sulfur vapor may be a reference to the bleaching ability of sulfur dioxide (produced from burning sulfur); newspaper is still bleached by that method today.

14. The reference is to the production of mercuric sulfide, which is a solid (unlike liquid mercury) and far less volatile than sulfur.

15. See Matteo Martelli, "'Divine Water' in the Alchemical Writings of Pseudo-Democritus," *Ambix* 56 (2009): 5–22, and Cristina Viano, "Gli alchimisti greci e l'acqua divina," *Rendiconti della Accademia Nazionale delle Scienze. Parte II: Memorie di scienze fisiche e naturali* 21 (1997): 61–70.

16. Mertens, *Les Alchimistes grecs IV, i: Zosime*, p. 21. On his use of "hermaphrodite," see chapter 3, note 69.

17. In an exciting recent development, several long-lost texts of Zosimos have been identified in Arabic translation. These documents, along with many others spuriously attributed to Zosimos, have been known for some time (Manfred Ullmann, *Die Natur- und Geheimwissenschaften im Islam* [Leiden: Brill, 1972], pp. 160–64), but their authenticity has only recently been established by Benjamin Hallum ("Zosimus Arabus," PhD diss., Warburg Institute, 2008). These will be edited and published in due course. I quote here from the "Twenty-Sixth Epistle," p. 366.

18. Mertens, *Les Alchimistes grecs IV, i: Zosime*, p. 17.

19. Ibid., pp. 40–41.

20. Ibid., p. 47.

21. Hallum, "Zosimus Arabus," pp. 130–47, quoting from pp. 142–43; compare with the interpretation given by Mertens, *Les Alchimistes grecs IV, i: Zosime*, p. 45, note 19. *On Sulphurs* may prove to be the only complete or nearly complete work of Zosimos that survives, and has already been shown to be the original source for two isolated fragments previously known in Greek.

22. Mertens, *Les Alchimistes grecs IV, i: Zosime*, pp. 207–31.

23. Ibid., p. 41.

24. On Gnosticism, see Wouter J. Hanegraaff, Antoine Faivre, Roelof van den Broek, and Jean-Pierre Brach, eds., *The Dictionary of Gnosis and Western Esotericism* (Leiden: Brill, 2005), 1:403–16 and references therein.

25. A good English translation of this prologue exists as Zosimos of Panopolis, *On the Letter Omega*, ed. and trans. Howard M. Jackson (Missoula, MT: Scholars Press, 1978); a more rigorously critical edition, with commentary, is in Mertens, *Les Alchimistes grecs IV, i: Zosime*, pp. 1–10. The "Final Account" is edited (with French translation) in Festugière, *Révélation*, pp. 275–81, 363–68. For further analysis see Daniel Stolzenberg, "Unpropitious Tinctures: Alchemy, Astrology, and Gnosis according to Zosimos of Panopolis," *Archives internationales d'histoire des sciences* 49 (1999): 3–31.

26. In classical thought, daimons are immaterial beings occupying an intermediate rank between the gods and men. Their moral proclivities might be either benign or wicked (Socrates referred to a daimon that gave him valuable advice), but in Zosimos's cosmological view they seem always to be intent on keeping men enslaved. His perspective may reflect the influence of Jewish and/or Christian ideas.

27. Zosimos, "Final Account," in Festugière, *Revelation*, p. 366.

28. *Acta sanctorum julii* (Antwerp, 1719–31), 2:557; John of Antioch, *Iohannes Antiocheni fragmenta ex Historia chronica*, ed. and trans. Umberto Roberto (Berlin: De Gruyter, 2005), fragment 248, pp. 428–29.

29. C. H. V. Sutherland, "Diocletian's Reform of the Coinage: A Chronological Note," *Journal of Roman Studies* 45 (1955): 116–18; Juan Carlos Martinez Oliva, "Monetary Integration in the Roman Empire," in *From the Athenian Tetradrachm to the Euro*, ed. P. L. Cottrell, Gérasimos Notaras, and Gabriel Tortella (Burlington, VT: Ashgate, 2007), pp. 7–23, esp. pp. 18–22.

30. Paul T. Keyser, "Greco-Roman Alchemy and Coins of Imitation Silver," *American Journal of Numismatics* 7–8 (1995): 209–33.

31. From a fragment of Zosimos cited in the ninth century by Georgos Synkellos, *Chronographia*, 1:23–24; for analysis, see Mertens, *Les Alchimistes grecs IV, i: Zosime*, pp. xciii–xcvi. We do not know the context in which Zosimos originally wrote this idea.

32. Plutarch, *De Iside et Osiride*, 33:364C.

33. Robert Halleux, *Les textes alchimiques* (Turnhout, Belgium: Brepols, 1979), pp. 45–47.

34. For an overview see Michèle Mertens, "Graeco-Egyptian Alchemy in Byzantium," in *The Occult Sciences in Byzantium*, ed. Paul Magdalino and Maria Mavroudi (Geneva: La Pomme d'Or, 2006), pp. 205–30.

35. Cristina Viano, "Les alchimistes gréco-alexandrins et le *Timée* de Platon," in *L'Alchimie et ses racines philosophiques: La tradition grecque et la tradition arabe*, ed. Cristina Viano (Paris: Vrin, 2005), pp. 91–108; "Aristote et l'alchimie grecque," *Revue d'histoire des sciences* 49 (1996): 189–213; *La matière des choses: Le livre IV des Météorologiques d'Aristote et son interprétation par Olympiodore* (Paris: Vrin, 2006), esp. appendix 1, pp. 199–208: "Olympiodore l'alchimiste"; "Olympiodore l'alchimiste et les Présocratiques," in *Alchemie: Art, histoire, et mythes*, ed. Didier Kahn and Sylvain Matton (Paris: SÉHA, 1995), pp. 95–150; and "Le commentaire d'Olympiodore au livre IV des *Météorologiques* d'Aristote," in *Aristoteles chemicus*, ed. Cristina Viano (Sankt Augustin, Germany: Academia Verlag, 2002), pp. 59–79.

36. There has been a long debate whether the Stephanos of the *Corpus alchemicum graecum* is the same person as the Neoplatonic philosopher Stephanos. The most recent evidence leads to the conclusion that he is. See Maria K. Papathanassiou, "L'Oeuvre alchimique de Stephanos d'Alexandrie," in Viano, *L'Alchimie et ses racines*, pp. 113–33; "Stephanus of Alexandria: On the Structure and Date of His Alchemical Work," *Medicina nei secoli* 8 (1996): 247–66; and "Stephanos of Alexandria: A Famous Byzantine Scholar, Alchemist and Astrologer," in Madgalino and Mavroudi, *Occult Sciences*, pp. 163–203. A rough English translation is available in Frank Sherwood Taylor, "Alchemical Works of Stephanus of Alexandria, Part I," *Ambix* 1 (1937): 116–39, and "Part II," *Ambix* 2 (1938): 39–49.

37. The phrase "stone that is no stone" appears in Zosimos (Mertens, *Les alchimistes grecs IV, i: Zosime*, p. 49). Note that the correct term is *Philosophers' Stone*, not the commonly encountered *Philosopher's Stone*. All original sources in various languages use the plural possessive: *Stone of the Philosophers*.

Chapter Two

1. Marcellin Berthelot, Rubens Duval, and O. Houdas, *La chimie au moyen âge*, 3 vols. (Paris, 1893).

2. For a good treatment of the translation movement, see Dimitri Gutas, *Greek Thought, Arabic Culture: The Graeco-Arabic Translation Movement in Baghdad and Early 'Abbasid Society* (London: Routledge, 1998). For a quicker introduction, see David C. Lindberg, *The Beginnings of Western Science*, 2nd ed. (Chicago: University of Chicago Press, 2007), pp. 166–76.

3. This information comes from the *Catalogue (al-Fihrist)* composed in 987 by the Baghdad bookseller Ibn al-Nadīm, one of the greatest resources for bibliographers of Arabic sources. An English translation of the section covering alchemy is J. W. Fück, "The Arabic Literature on Alchemy according to An-Nadīm," *Ambix* 4 (1951): 81–144; this section contains an early version of the story of Khālid and his books on p. 89 and in the notes on p. 120.

4. Morienus, *De compositione alchemiae*, in *Bibliotheca chemica curiosa*, ed. J. J. Manget (Geneva, 1702; reprint, Sala Bolognese: Arnoldo Forni, 1976), 1:509–19; Ullmann, *Natur- und Geheimwissenschaften*, pp. 191–95; Ahmad Y. al-Hassan, "The

Arabic Original of the *Liber de compositione alchemiae,*" *Arabic Sciences and Philosophy* 14 (2004): 213–31.

5. Julius Ruska, *Arabische Alchemisten I: Chālid ibn-Jazīd ibn-Mu'āwija, Heidelberger Akten von-Portheim-Stiftung* 6 (1924; reprint, Vaduz, Liechtenstein: Sändig Reprint Verlag, 1977); Manfred Ullmann, "Hālid ibn-Yazīd und die Alchemie: Eine Legende," *Der Islam* 55 (1978): 181–218.

6. For example, Patriarch Timothy I prepared the first Arabic version of a work by Aristotle (the *Topics*) for the caliph al-Mahdī around 782; Gutas, *Greek Thought,* pp. 61–69.

7. For short descriptions of these early productions, see Georges C. Anawati, "L'alchimie arabe," in *Histoire des sciences arabes,* ed. Roshdi Rashed and Régis Morelon, vol. 3, *Technologie, alchimie et sciences de la vie* (Paris: Seuil, 1997), pp. 111–42, and Ullmann, *Natur- und Geheimwissenschaften,* pp. 151–91.

8. On Hermes and Hermeticism, see Hanegraaff, Faivre, van den Broek, and Brach, *Dictionary of Gnosis and Western Esotericism,* 1:474–570; Garth Fowden, *The Egyptian Hermes: A Historical Approach to the Late Pagan Mind* (Cambridge: Cambridge University Press, 1986) [useful for Hermes, but the material on Zosimos and alchemy is now outdated]; and Florian Ebeling, *The Secret History of Hermes Trismegestus: Hermeticism from Ancient to Modern Times* (Ithaca, NY: Cornell University Press, 2007), pp. 3–36; for the philosophico-theological texts see Brian Copenhaver, *Hermetica: The Greek Corpus Hermeticum and the Latin Asclepius* (Cambridge: Cambridge University Press, 1992).

9. For the Arabic Hermes and texts attributed to him, see Ullmann, *Natur- und Geheimwissenschaften,* pp. 165–72 and 368–78; Fück, "An-Nadim," pp. 89–91; and Martin Plessner, "Hermes Trismegistus and Arab Science," *Studia Islamica* 2 (1954): 45–59. On the growth of the Arabic myth of Hermes (with little about alchemy), see Kevin T. Van Bladel, *The Arabic Hermes: From Pagan Sage to Prophet of Science* (Oxford: Oxford University Press, 2009).

10. For one version of Hermes Trismegestus as the ancient father of alchemy, see Michael Maier, *Symbola aureae mensae duodecim nationum* (Frankfurt, 1617), pp. 5–19; for the usage of the term *Hermetic Art,* see Bernard Joly, "La rationalité de l'Hermétisme: La figure d'Hermès dans l'alchimie à l'âge classique," *Methodos* 3 (2003): 61–82, and Jean Beguin, *Tyrocinium chymicum* (Paris, 1612), pp. 1–2: "If anyone should call it [chymistry] the Hermetic Art, he refers to its originator and antiquity." For the seventeenth-century attack on Hermes' date and prophetic status, see Anthony Grafton, "Protestant versus Prophet: Isaac Casaubon on Hermes Trismegistus," *Journal of the Warburg and Courtauld Institutes* 46 (1983): 78–93. For a lengthy early modern alchemical commentary, see Gerhard Dorn, *Physica Trismegesti,* in *Theatrum chemicum,* 1:362–87; and on Newton, J. E. McGuire and P. M. Rattansi, "Newton and the Pipes of Pan," *Notes and Records of the Royal Society of London* 21 (1966): 108–43, and B. J. T. Dobbs, "Newton's Commentary on *The Emerald Tablet* of Hermes Trismegestus: Its Scientific and Theological Significance," in *Hermeticism and the Renaissance,* ed. Ingrid Merkel and Allen G. Debus (Washington, DC: Folger Shakespeare Library, 1988), pp. 182–91.

11. Julius Ruska, *Tabula Smaragdina: Ein Beitrag zur Geschichte der hermetischen Literatur* (Heidelberg: Winter, 1926); Martin Plessner, "Neue Materialien zur Geschichte der Tabula Smaragdina," *Der Islam* 16 (1928): 77–113; for a summary of the

history of the *Tablet* and several versions of its text, see Didier Kahn, ed., *La table d'émeraude et sa tradition alchimique* (Paris: Belles Lettres, 1994).

12. Balīnūs is actually the Arabic rendering of Apollonios. Arabic has no *p*, so that letter becomes *b*, giving "Abollonios," which, with the vowel modifications typical of translating into and out of Arabic (the language has only three vowels—*a*, *i*, and *u*—and does not indicate short vowels in writing), gives "Balīnūs."

13. The Arabic text of the *Kitāb sirr al-khalīqa* was not edited and published until 1979: Ursula Weisser, ed., *Sirr al-khalīqah wa ṣan'āt al-ṭabī'ah* (Aleppo: Aleppo Institute for the History of Arabic Science, 1979). A summary of its contents is currently available in Ursula Weisser, *Das "Buch über das Geheimnis der Schöpfung" von Pseudo-Apollonios von Tyana* (Berlin: Walter de Gruyter, 1980; reprint, 2010), and an edition of the medieval Latin translation edited by Françoise Hudry is "Le *De secretis naturae* du pseudo-Apollonius de Tyane: Traduction latine par Hugues de Santalla du *Kitāb sirr al-ḫalīqa* de Balīnūs," in "Cinq traités alchimique médiévaux," *Chrysopoeia* 6 (1997–99): 1–153.

14. The discovery of arcane texts in underground burial chambers or in ancient Egyptian monuments had become a literary device by the Islamic period; see Ruska, *Tabula*, pp. 61–68.

15. English translation from the Arabic provided by E. J. Holmyard, "The Emerald Table," *Nature* 112 (1923): 525–26, quoting from p. 526. Note, however, that Holmyard's historical claims in this article about the origin and dating of the *Tabula* have since been shown to be wrong.

16. Gotthard Strohmaier, "'Umāra ibn Hamza, Constantine V, and the Invention of the Elixir," *Graeco-Arabica* 4 (1991): 21–24; a fuller account is Strohmaier, "Al-Mansūr und die frühe Rezeption der griechischen Alchemie," *Zeitschrift für Geschichte der Arabisch-Islamischen Wissenschaften* 5 (1989): 167–77.

17. Fück, "An-Nadim," p. 96.

18. Paul Kraus, *Jābir ibn Ḥayyān: Contribution à l'histoire des idées scientifiques dans l'Islam*, vol. 1, *Le Corpus des écrits jābiriens*, Mémoires de L'Institut d'Égypte 44 (1943), and vol. 2, *Jābir et la science grecque*, Mémoires de L'Institut d'Égypte 45 (1942). The second volume has been reprinted by Les Belles Lettres (Paris, 1986). In 1944, as Kraus was completing a third book that situated Jābir in the context of Islamic religious history, he was found hanged in his apartment in Cairo. Doubts remain whether it was suicide or murder. To make matters worse, much of the manuscript of his third book was lost after his death. This unequaled scholar who solved so many difficult mysteries of the past departed tragically in a mystery of his own; and while he painstakingly recovered books that had been lost for centuries, most of his own final book perished due to carelessness.

19. An alchemical text bearing the name of Ja'far has been found, but it has been shown to be a later forgery. See Julius Ruska, *Arabische Alchemisten II: Ǧa'far alṣādiq, der Sechste Imām, Heidelberger Akten von-Portheim-Stiftung* 10 (1924; reprint, Vaduz, Liechtenstein: Sändig Reprint Verlag, 1977). This publication includes a German translation of the alchemical text attributed to Ja'far.

20. This view is summarized in Kraus, *Le Corpus des écrits jābiriens*, pp. xlv–lxv.

21. Aristotle, *Meteorologica* 3.6.378a17–b6.

22. Kraus, *Jābir et la science grecque*, pp. 270–303, and Pinella Travaglia, "I *Meteorologica* nella tradizione eremetica araba: il *Kitāb sirr al-ḫaliqa*," in Viano, *Aristoteles chemicus*, pp. 99–112.

23. Jābir, *Kitāb al-īḍāḥ*, in *The Arabic Works of Jābir ibn Ḥayyān*, ed. and trans. E. J. Holmyard (Paris: Geuthner, 1928), p. 54 [Arabic text]; E. J. Holmyard, "Jābir ibn-Ḥayyān," *Proceedings of the Royal Society of Medicine, Section of the History of Medicine* 16 (1923): 46–57, quoting from p. 56 [partial English translation]; Karl Garbers and Jost Weyer, eds., *Quellengeschichtliches Lesebuch zur Chemie und Alchemie der Araber im Mittelalter* (Hamburg: Helmut Buske Verlag, 1980), pp. 34–35 [German and Arabic].

24. Greek alchemists did not classify mercury as one of the metals; some texts in the Jābirian corpus do and others do not. It was generally considered a metal in later Arabic alchemy and in Latin alchemy.

25. Garbers and Weyer, *Lesebuch*, pp. 14–15; Holmyard, "Jābir," p. 57.

26. See Lindberg, *Beginnings of Western Science*, pp. 31, 53–54.

27. These processes are described in detail in Kraus, *Jābir et la science grecque*, pp. 4–18.

28. For a quick overview of Galenic medicine, see G. E. R. Lloyd, *Greek Science after Aristotle* (New York: Norton, 1973), pp. 136–53, esp. 138–40; for the development of the system of degrees by al-Kindī, see Pinella Travaglia, *Magic, Causality and Intentionality: The Doctrine of Rays in al-Kindī*, Micrologus Library 3 (Florence: Sismel, 1999), pp. 73–96.

29. Indeed, Jābir says that the opposite qualities already exist at the "interior" of substances, and so need to be exchanged with the contrary exterior properties. See Kraus, *Jābir et la science grecque*, pp. 1–3.

30. Very little of the Jābirian corpus has been published. A selection of edited Arabic texts exists in Holmyard, *The Arabic Works of Jābir ibn-Ḥayyān*; Paul Kraus, *Jābir ibn-Ḥayyān: Textes choisis* (Paris: Maisonneuve, 1935); and Pierre Lory, *L'Élaboration de l'Élixir Suprême* (Damascus: Institut Français de Damas, 1988) [the first fourteen treatises from the *Seventy Books*]. Translations into European languages (none into English) are Alfred Siggel, ed., *Das Buch der Gifte des Ğābir ibn-Ḥayyān* (Wiesbaden: Akademie der Wissenschaften und der Literatur, 1958) [Arabic of *Kitāb al-sumūm*, with German translation], and Pierre Lory, trans., *Dix traités d'alchimie* (Paris: Sinbad, 1983) [first ten treatises of the *Seventy Books* into French]. The earliest text, the *Kitāb al-raḥma*, also exists in a medieval Latin translation, first edited by Ernst Darmstaedter, "Liber Misericordiae Geber: Eine lateinische Übersetzung des grösseren Kitāb alrahma," *Archiv für Geschichte der Medizin* 17 (1925): 187–97, and a medieval Latin translation of the *Seventy Books* is published by Marcellin Berthelot, *Mémoires de l'Académie des Sciences* 49 (1906): 308–77. Lory (*Dix traités*, pp. 79–89) has a good account of Jābir's apparatus and operations, and Kraus (*Jābir et la science grecque*, pp. 3–18) lays out the steps in the preparation of the elixirs very clearly, with extended practical passages rendered into French, pp. 3–18.

31. Kraus, *Jābir et la science grecque*, pp. 6–7; Lory, *Dix traités*, pp. 91–94.

32. For an entrée into Pythagoreanism, see Carl Huffman's article in Jacques Brunschwig and Geoffrey E. R. Lloyd, eds., *Greek Thought: A Guide to Classical Knowledge* (Cambridge, MA: Belknap Press of Harvard University Press, 2000), pp. 918–36; for a valuable overview of number symbolism, see Jean-Pierre Brach's article "Number Symbolism" in Hanegraaff, Faivre, van den Broek, and Brach, *Dictionary of Gnosis and Western Esotericism*, 2:874–83.

33. John 21:3–14. The early church fathers, formed in the intellectual culture of late antiquity, had no problem "reading" this number of fish as one of completion and universality, the catch of 153 fish meaning every race and nation on earth would be

saved within the church, the net that did not break despite the strain; see for example St. Augustine, *On the Gospel of John*, tractate 122. To understand a "triangular number," draw a single dot, then two dots in a row beneath it to mark the corners of an equilateral triangle. Expand the triangle with a row of three dots beneath the row of two, and continue with rows of four, five, six, and so forth. When you reach seventeen rows, the total number of dots will be 153, the "triangular number" of 17.

34. Jābir, quoted in Kraus, *Le corpus des écrits jābiriens*, p. xxvii.

35. Ibid.

36. Ibid., pp. xxxiii–xxxiv.

37. Julius Ruska and E. Wiedemann catalog some Arabic *Decknamen* found in a work of al-Tughrā'ī (eleventh century) in "Beiträge zur Geschichte der Naturwissenschaften LXVII: Alchemistische Decknamen," *Sitzungsberichte der Physikalisch-medizinalischen Societät zu Erlangen* 56 (1924):17–36; a longer list drawn from more sources is given by Alfred Siggel, *Decknamen in der arabischen alchemistischen Literatur* (Berlin: Akademie Verlag, 1951).

38. Quoted from the *Book of Properties* (*Kitāb al-khawāss*) in Kraus, *Le corpus des écrits jābiriens*, p. xxviii.

39. William R. Newman, *The Summa Perfectionis of the Pseudo-Geber: A Critical Edition, Translation, and Study* (Leiden: Brill, 1991), p. 90.

40. Robert Boyle, *Dialogue on Transmutation*, edited in Lawrence M. Principe, *The Aspiring Adept: Robert Boyle and His Alchemical Quest* (Princeton, NJ: Princeton University Press, 1998), pp. 233–95, quoting from pp. 273–74; text modernized here.

41. Martin Plessner, "The Place of the *Turba Philosophorum* in the Development of Alchemy," *Isis* 45 (1954): 331–38, and *Vorsokratische Philosophie und griechische Alchemie* (Wiesbaden: Steiner, 1975). Plessner's work extends and corrects the foundational, and still valuable, work on the text by Julius Ruska, *Turba philosophorum: Ein Beitrag zur Geschichte der Alchemie* (Berlin: Springer, 1931).

42. Julius Ruska, "Al-Biruni als Quelle für das Leben und die Schriften al-Rāzī's," *Isis* 5 (1923): 26–50; "Die Alchemie ar-Razi's," *Der Islam* 22 (1935): 281–319.

43. Julius Ruska, *Al-Rāzī's Buch der Geheimnis der Geheimnisse* (Berlin: Springer, 1937; reprint, Graz: Verlag Geheimes Wissen, 2007) [contains a complete German translation of Al-Rāzī's text]; H. E. Stapleton, R. F. Azo, and M. Hidayat Husain, "Chemistry in Iraq and Persia in the Tenth Century AD," *Memoirs of the Asiatic Society of Bengal* 8 (1927): 317–418 [contains a partial English translation of Al-Rāzī's text].

44. There is a brief comment by the fifth-century Neoplatonic philosopher Proclus that seems to deny that alchemists can make gold in the same way that nature does, although it is not clear that he denies chrysopoeia itself; Proclus, *Commentary on the Republic*, 2.234.17.

45. On al-Kindī, see Felix Klein-Francke, "Al-Kindi," in *The History of Islamic Philosophy*, ed. Seyyed Hossein Nasr and Oliver Leaman (New York: Routledge, 1996), pp. 165–77. Reference to his lost work against chrysopoeia is made by al-Mas'ūdī (died 956) in his *Murūj al-dhahab*, available in French translation: *Les Prairies d'Or*, trans. B. de Maynard and P. de Courteille (Paris, 1861–1917), 5:159.

46. The work is listed in medieval catalogs (both Arabic and Latin) of al-Razi's works, see G. S. A. Ranking, "The Life and Works of Rhazes (Abu Bakr Muhammad bin Zakariya ar-Razi)," *XVII International Congress of Medicine, London 1913, Proceedings*, section 23, pp. 237–68; on p. 249, no. 40.

47. Julius Ruska, "Die Alchemie des Avicenna," *Isis* 21 (1934): 14–51, judged the work to be a Latin forgery, but an Arabic text exists; see H. E. Stapleton, R. F. Azo, Hidayat Husain, and G. L. Lewis, "Two Alchemical Treatises Attributed to Avicenna," *Ambix* 10 (1962): 41–82. The Arabic text, a French translation, and the medieval Latin version are all provided in Georges C. Anawati, "Avicenna et l'alchimie," in *Convegno internazionale, 9–15 aprile 1969: Oriente e occidente nel medioevo: filosofia e scienze* (Rome: Accademia Nazionale dei Lincei, 1971), pp. 285–345.

48. E. J. Holmyard and D. C. Mandeville, eds., *Avicennae de congelatione et conglutinatione lapidum, Being Sections of the Kitāb al-Shifā'* (Paris: Paul Geuthner, 1927), p. 40. This edition contains Latin and Arabic texts with an English translation of the latter, plus notes.

49. Ibid., p. 41.

50. Ibn-Sīnā, quoted in A. F. Mehrens, "Vues d'Avicenne sur astrologie et sur le rapport de la responsabilité humaine avec le destin," *Muséon* 3 (1884): 383–403, quoting from p. 387.

51. Ibn-Sīnā, quoted in Holmyard and Mandeville, *Avicennae de Congelatione*, p. 41.

52. For a summary, see Ullmann, *Natur- und Geheimwissenschaften*, pp. 249–55.

53. One exception is an account of John Isthmeos, who appeared in Antioch in 504, swindling many people there before moving to Constantinople, where he continued his trade until he was exiled; see Mertens, "Graeco-Egyptian Alchemy," pp. 226–27.

54. The text exists in French translation as al-Jawbari, *La voile arraché*, trans. René R. Khawan, 2 vols. (Paris: Phèbus, 1979); the section on chrysopoeia is 1:183–229. A partial English translation appears in Harold J. Abrahams, "Al-Jawbari on False Alchemists," *Ambix* 31 (1984): 84–87.

55. Leo Africanus, *A Geographicall Historie of Africa* (London, 1600), pp. 155–56. The text was originally published in 1526 in Italian. On Fez as a continuing center of alchemy, see José Rodríguez Guerrero, "Some Forgotten Fez Alchemists and the Loss of the Peñon de Vélez de la Gomera in the Sixteenth Century," in *Chymia: Science and Nature in Medieval and Early Modern Europe*, ed. Miguel López-Pérez, Didier Kahn, and Mar Rey Bueno (Newcastle-upon-Tyne: Cambridge Scholars Publishing, 2010), pp. 291–309.

56. For a summary of some of these later alchemical authors, see Ullmann, *Natur- und Geheimwissenschaften*, pp. 224–48.

57. Holmyard, *Alchemy*, p. 104.

Chapter Three

1. Morienus, *De compositione alchemiae*, in *Bibliotheca chemica curiosa*, 1:509–19, quoting from p. 509; this Latin edition is fairly corrupt—I have silently changed its *vestra* to *nostra* in accord with some manuscripts. For an English translation and an alternate Latin text (omitting the prologue), see Morienus, *A Testament of Alchemy*, ed. and trans. Lee Stavenhagen (Hanover, NH: Brandeis University Press, 1974); the translation is not always accurate. The authenticity of the work as a translation from Arabic rather than as an original Latin composition was denied by Julius Ruska, *Arabische Alchemisten I*, pp. 33-35, but partial Arabic versions have since been found: Ullmann, *Natur- und Geheimwissenschaften*, pp. 192–93, and al-Hassan, "The Arabic Original." The authenticity of Robert's prologue as a twelfth-century work has also been called

into question, including by Stavenhagen (pp. 52–60), but has been reaffirmed convincingly by Richard Lemay, "L'authenticité de la Préface de Robert de Chester à sa traduction du *Morienus*," *Chrysopoeia* 4 (1990–91): 3–32; see also Didier Kahn, "Note sur deux manuscrits du Prologue attribué à Robert de Chester," ibid., pp. 33–34. A thorough critical edition of the Morienus text remains a desideratum. Of course, my citation of an exact day of the week on which alchemy "arrived" in the Latin world is partly tongue-in-cheek; there were undoubtedly some earlier transfers and multiple points of ingress. Nonetheless, the fact remains that we can trace the origins of Latin alchemy more clearly than those of Greek or Arabic alchemy.

2. The classic source is Charles Homer Haskins, *The Renaissance of the Twelfth Century* (Cambridge, MA: Harvard University Press, 1927); more recently, Robert L. Benson and Giles Constable, eds., *Renaissance and Renewal in the Twelfth Century*, with Carol D. Lanham (Cambridge, MA: Harvard University Press, 1982; reprint, Toronto: Medieval Academy of America, 1991); in regard to the Latin translation movement, see the article by Marie-Thérèse d'Alverny, "Translations and Translators," on pp. 421–62; see also Edward Grant, *The Foundations of Modern Science in the Middle Ages* (Cambridge: Cambridge University Press, 1996), pp. 18–32.

3. Morienus, *De compositione*, in *Bibliotheca chemica curiosa*, 1:509.

4. Hugh of Santalla's twelfth-century translation of Balīnūs is edited in Hudry, "Le *De secretis naturae*."

5. Cyril Stanley Smith and John G. Hawthorne, *Mappae Clavicula: A Little Key to the World of Medieval Techniques*, Transactions of the American Philosophical Society 64 (Philadelphia: American Philosophical Society, 1974); Rozelle Parker Johnson, *Compositiones variae: An Introductory Study*, Illinois Studies in Language and Literature 23 (Urbana, IL, 1939); Heinz Roosen-Runge, *Farbgebung und Technik frümittelalterlicher Buchmalerei: Studien zu den Traktaten "Mappae Clavicula" und "Heraclius,"* 2 vols. (Munich: Deutscher Kunstverlag, 1967).

6. Theophilus is possibly identifiable as Roger of Helmarshausen, a Benedictine monk; his book in available in English translation as *On Divers Arts*, trans. John G. Hawthorne and Cyril Stanley Smith (New York: Dover, 1979).

7. Ibid., pp. 119–20.

8. Carmélia Opsomer and Robert Halleux, "L'Alchimie de Théophile et l'abbaye de Stavelot," in *Comprendre et maîtriser la nature au Moyen Age*, ed. Guy Beaujouan (Geneva: Droz, 1994), pp. 437–59, and Halleux, "La réception de l'alchimie arabe en Occident," in Rashed and Morelon, *Histoire des sciences arabes*, 3:143–51, esp. pp. 143–45.

9. One of the earliest of these is the *Ars alchemie*, dating from the early thirteenth century; see Antony Vinciguerra, "The *Ars alchemie*: The First Latin Text on Practical Alchemy," *Ambix* 56 (2009): 57–67.

10. We owe this identification, and the solution to the "Jābir-Geber" problem, to the painstaking studies of William R. Newman. For a detailed treatment of Geber's identity, see Newman, "New Light on the Identity of Geber," *Sudhoffs Archiv* 69 (1985): 79–90, and "Genesis of the *Summa perfectionis*," *Archives internationales d'histoire des sciences* 35 (1985): 240–302. For an edition, translation, and historical contextualization of the *Summa*, see Newman's *The Summa Perfectionis of Pseudo-Geber*.

11. On such borrowings from Jābir, see Newman, *Summa perfectionis*, pp. 86–99.

12. One notable exception to this generalization is Roger Bacon, who was apparently more deeply influenced by Jābir than were others; see William R. Newman, "The Philosophers' Egg: Theory and Practice in the Alchemy of Roger Bacon," in "Le crisi

dell'alchimia," *Micrologus* 3 (1995): 75–101, and Michela Pereira, "Teorie dell'elixir nell'alchimia latina medievale," in ibid., pp. 103–49.

13. Aristotle, *Physics* 187b14–22, and *Meteors* 385b12–26, 386b1–10 and 387a17–22; for the expansion of these ideas in the Middle Ages, especially in relation to Geber, see Newman, *Summa perfectionis*, pp. 167–90. On *Meteors IV* and its importance to alchemy, see the essays in Viano, *Aristoteles chemicus*, and Craig Martin, "Alchemy and the Renaissance Commentary Tradition on *Meteorologica* IV," *Ambix* 51 (2004): 245–62.

14. Newman, *Summa perfectionis*, pp. 159–62, 471–75, and 725–26.

15. Ibid., pp. 143–92; William R. Newman, *Atoms and Alchemy* (Chicago: University of Chicago Press, 2006), esp. pp. 23–44; Antoine Calvet, "La théorie *per minima* dans les textes alchimiques des XIVe et XVe siècles," in *Chymia: Science and Nature in Medieval and Early Modern Europe*, ed. Miguel López-Pérez, Didier Kahn, and Mar Rey Bueno (Newcastle-upon-Tyne: Cambridge Scholars Publishing, 2010), pp. 41–69.

16. There are many minor variants of the Latin text; see Newman, *Summa perfectionis*, pp. 48–51. Holmyard and Mandeville, *Avicennae de congelatione*, pp. 53–54, gives one version; another is *Avicennae de congelatione et conglutinatione lapidum*, in *Bibliotheca chemica curiosa*, pp. 636–38, quotation from p. 638. In fact, Aristotle actually had a much higher regard for the power of human artifice than did Ibn-Sīnā.

17. Attention was first called to the manuscript by Newman, who published and analyzed a portion of it. See Newman, *Summa perfectionis*, pp. 7–15.

18. On Albert's alchemy, see Pearl Kibre, "Albertus Magnus on Alchemy," in *Albertus Magnus and the Sciences: Commemorative Essays 1980*, ed. James A. Weisheipl (Toronto: Pontifical Institute of Mediaeval Studies, 1980), pp. 187–202; "Alchemical Writings Attributed to Albertus Magnus," *Speculum* 17 (1942): 511–15; and Robert Halleux, "Albert le Grand et l'alchimie," *Revue des sciences philosophiques et théologiques* 66 (1982): 57–80. For his own works on alchemy, *Liber mineralium*, in *Alberti Magni opera omnia*, ed. A. Borgnet (Paris, 1890–99), 5:1–116, and the attributed *Libellus de alchemia*, 37:545–73; English translations: *Book of Minerals*, trans. Dorothy Wyckoff (Oxford: Clarendon Press, 1967), and *"Libellus de Alchemia" Ascribed to Albertus Magnus*, trans. Virginia Heines, SCN (Berkeley: University of California Press, 1958).

19. St. Thomas Aquinas, *Summa theologica*, 2ae 2a, quaestio 77, articulus 2.

20. Giles of Rome, *Quodlibeta*, quaestio 3, quolibet 8, in Sylvain Matton, *Scolastique et Alchimie*, Textes et Travaux de Chrysopoeia 10 (Paris: SÉHA; Milan: Arché, 2009), pp. 77–80; William R. Newman, "Technology and Alchemical Debate in the Late Middle Ages," *Isis* 80 (1989): 423–45, esp. pp. 437–39.

21. *Libellus*, trans. by Heines, p. 19; St. Albert, *Book of Minerals*, p. 179. It is possibly from his teacher St. Albert that Thomas Aquinas took the notion that alchemical gold has different properties from natural gold.

22. William R. Newman explores the connection between technology and alchemy, art and nature more fully in his fascinating and provocative *Promethean Ambitions: Alchemy and the Quest to Perfect Nature* (Chicago: University of Chicago Press, 2004).

23. Aristotle himself could be invoked to support this position, for he wrote that "art completes whatever nature is unable to complete": *Physics* 2.8; 199a 15–16.

24. Reported by Nicholas Eymerich in 1396. See Halleux, *Les textes alchimiques*, p. 126.

25. The full text of *Spondent quas non exhibent* (*They Promise What They Do Not Deliver*) is printed in ibid., pp. 124–26, in original Latin and a French translation.

26. Ibid., p. 124.

27. For Henry IV in 1404 (5 Hen. 4), see A. Luders et al., eds., *The Statutes of the Realm* (London, 1816), 2:144; for Venetian Council of Ten in 1488, see Pantheus, *Voarchadumia*, in *Theatrum chemicum* (Strasbourg, 1659–63), 2:495–549, on pp. 498–99.

28. Ibn-Khaldūn, *The Muqaddimah: An Introduction to History* (New York: Pantheon, 1958), 3:277.

29. A few later examples are Johannes Chrysippus Fanianus, *De jure artis alchimiae*, in *Theatrum chemicum*, 1:48–63; Girolamo de Zanetis, *Conclusio*, in ibid., 4:247–52; and Johann Franz Buddeus, *Quaestionem politicam an alchimistae sint in republica tolerandi?* (Magdeburg, 1702), in German translation as *Untersuchung von der Alchemie*, in *Deutsches Theatrum Chemicum*, ed. Friedrich Roth-Scholtz (Nuremberg, 1728), 1:1–146. For discussion of the topic, see Ku-ming (Kevin) Chang, "Toleration of Alchemists as a Political Question: Transmutation, Disputation, and Early Modern Scholarship on Alchemy," *Ambix* 54 (2007): 245–73, and Jean-Pierre Baud, *Le procès d'alchimie* (Strasbourg: CERDIC, 1983).

30. D. Geoghegan, "A Licence of Henry VI to Practise Alchemy," *Ambix* 6 (1957): 10–17.

31. Eilhard Wiedemann, "Zur Alchemie bei der Arabern," *Journal für Praktische Chemie* 184 (1907): 115–23 provides a German translation of al-Fārābī's work.

32. For example, we have a text dating from 1257 that appears to represent university lectures that include alchemical knowledge; Constantine of Pisa, *The Book of the Secrets of Alchemy*, ed. and trans. Barbara Obrist (Leiden: Brill, 1990). Paul of Taranto was himself a lecturer in a Franciscan school.

33. For the latest publication on John of Rupescissa in English, see Leah DeVun, *Prophecy, Alchemy, and the End of Time: John of Rupescissa in the Late Middle Ages* (New York: Columbia University Press, 2009). Older but more exhaustive sources are Jeanne Bignami-Odier, "Jean de Roquetaillade," in *Histoire litteraire de la France* (Paris: Academie des Inscriptions et Belles-Lettres, 1981), 41:75–240, and Robert Halleux, "Ouvrages alchimiques de Jean de Rupescissa," in ibid., 41:241–77.

34. See David Burr, *The Spiritual Franciscans: From Protest to Persecution in the Century after St. Francis* (University Park: Penn State University Press, 2001).

35. John's text appears under two different titles: John of Rupescissa, *Liber lucis*, in *Bibliotheca chemica curiosa*, 2:84–87, and *De confectione veri lapidis philosophorum*, in ibid., 2:80–83. The two differ in details of wording and also the opening and closing text, but share the same structure, order, ideas, and practical details; the relationship between the two versions remains unresolved. The quotation used here is from the prologue (*Liber lucis*, 2:84), which is lacking in *De confectione*.

36. Rupescissa, *De confectione*, 2:83. Because we lack a critical edition of John's work, I hesitate to assert that the paragraphs about salt are his; they might have been added by a later follower who recognized the need for salt. These sections are absent from the *Liber lucis*.

37. For more on this topic, see chapter 6 and Lawrence M. Principe, "Chemical Translation and the Role of Impurities in Alchemy: Examples from Basil Valentine's *Triumph-Wagen*," *Ambix* 34 (1987): 21–30.

38. Rupescissa, *De confectione*, 2:81; the corresponding version in *Liber lucis* (2:84) is unclear, and may result from the loss of a line of text by a copyist. John is correct in

his observation about the weight gain; we now know that the mercury has combined with the chlorine of the salt, adding to the overall weight of the sublimed mercuric chloride.

39. On the interpenetration of medicine and Christianity in Arnald's genuine writings, analogous to that found with the alchemy of pseudo-Arnaldian texts, see Joseph Ziegler, *Medicine and Religion c. 1300: The Case of Arnau de Vilanova* (Oxford: Clarendon Press, 1998); a useful biographical sketch is on pp. 21–34. See also Chiara Crisciani, "Exemplum Christi e sapere: Sull'epistemologia di Arnoldo da Villanova," *Archives internationales d'histoire des sciences* 28 (1978): 245–87, and Antoine Calvet, "Alchimie et Joachimisme dans les *alchimica* pseudo-Arnaldiens," in *Alchimie et philosophie à la Renaissance*, ed. Jean-Claude Margolin and Sylvain Matton (Paris: Vrin, 1993), pp. 93–107.

40. Pseudo-Arnald of Villanova, *Tractatus parabolicus*, ed. and trans. [into French] Antoine Calvet, *Chrysopoeia* 5 (1992–96): 145–71. For an analysis, see Antoine Calvet, "Un commentaire alchimique du XIVᵉ siècle: Le *Tractatus parabolicus* du ps.-Arnaud de Villaneuve," in *Le Commentaire: Entre tradition et innovation*, ed. Marie-Odile Goulet-Cazé (Paris: Vrin, 2000), pp. 465–74. See also Antoine Calvet, *Les Oeuvres alchimiques attribuées à Arnaud de Villaneuve*, Textes et Travaux de Chrysopoeia 11 (Paris: SÉHA; Milan: Archè, 2011).

41. Pseudo-Arnald, *Tractatus*, p. 160.

42. "Le but poursuivi par l'auteur serait en somme d'asseoir l'alchimie sur un roc afin de confondre ses détracteurs"; Calvet, "Commentaire," p. 471.

43. Petrus Bonus, *Margarita pretiosa novella*, in *Bibliotheca chemica curiosa*, 2:1–80, quoting from pp. 30 and 50.

44. Rupescissa, *De confectione*, 2:81–82.

45. Rupescissa, *Liber lucis*, 2:85.

46. Rupescissa, *De confectione*, 2:81.

47. Upon heating the mixture, assuming common salt is also present, the mercury is converted into solid mercuric chloride. *Argentum vivum* is the source for our own alternate name for mercury, namely "quicksilver," where *quick* carries the archaic English meaning of *alive*.

48. While we lack an easily accessible edition or reprint of the *De consideratione*, there are three early printings: Basel, 1561(?) and 1597, and Ursel, 1602 (in *Theatrum chemicum*, 3:359–485; not present in the later editions). A fifteenth-century English edition is published as *The Book of the Quinte Essence*, ed. F. J. Furnivall (London: Early English Text Society, 1866; reprint, Oxford: Oxford University Press, 1965). A useful overview of the book's contents appears in Halleux, "Ouvrages alchimiques," pp. 245–62, and in Udo Benzenhöfer, *Johannes' de Rupescissa Liber de consideratione quintae essentiae omnium rerum deutsch* (Stuttgart: Franz Steiner Verlag, 1989), pp. 15–21. The latter contains an edition of a fifteenth-century German version of the text. See also Giancarlo Zanier, "Procedimenti farmacologici e pratiche chemioterapeutiche nel *De consideratione quintae essentiae*," in "Alchimia e medicina nel Medioevo," ed. Chiara Crisciani and Agostino Paravicini Bagliani, Micrologus Library 9 (Florence: Sismel, 2003), pp. 161–76.

49. Halleux, "Ouvrages alchimiques," pp. 246–50.

50. There had been several earlier claims that alchemy was useful for medicine, for example by Bernard of Gordon (died circa 1320); see Luke Demaitre, *Doctor Bernard de Gordon: Professor and Practitioner* (Toronto: Pontifical Institute of Medieval

Studies, 1980), pp. 19–20. Roger Bacon wrote that the Philosophers' Stone had medicinal virtues; see Michela Pereira, "Un tesoro inestimabile: Elixir e *prolongatio vitae* nell'alchimiae del '300," *Micrologus* 1 (1992): 161–87, and "Teorie dell'elixir."

51. For more on the linkages of alchemy and medicine in the Middle Ages, including some prior to John of Rupescissa, see the essays in Crisciani and Bagliani, "Alchimia e medicina nel Medioevo."

52. Michela Pereira, *The Alchemical Corpus Attributed to Raymond Lull* (London: Warburg Institute, 1989); "Sulla tradizione testuale del *Liber de secretis naturae seu de quinta essentia* attribuito a Raimondo Lullo," *Archives internationales d'histoire des sciences* 36 (1986): 1–16; "*Medicina* in the Alchemical Writings Attributed to Raimond Lull," in *Alchemy and Chemistry in the Sixteenth and Seventeenth Centuries*, ed. Piyo Rattansi and Antonio Clericuzio (Dordrecht: Kluwer, 1994), pp. 1–15.

53. Michela Pereira and Barbara Spaggiari, *Il Testamentum alchemico attribuito a Raimondo Lullo* (Florence: Sismel, 1999), contains critical editions of the Catalan original and a fifteenth-century Latin translation, plus valuable introductory materials.

54. Pseudo-Lull, *Testamentum* 2:1 and 3:7–10, in ibid., pp. 306–7 and 390–97; making precious stones was covered more fully by the same (anonymous) author in his *Liber lapidarius*. The *Book of the Secrets of Nature* contains the same threefold goals of alchemy, which stands to reason, since its author claims (falsely) that he is also the author of the *Testament*.

55. Ibid., 2:30, pp. 376–79.

56. This modern misconception comes possibly from a conflation of Chinese alchemical notions with European ones. Nevertheless, there are a *few* claims in the West of alchemists living to extreme ages. The story of the Nicolas and Pernelle Flamel living to over four hundred by use of the stone (reprised in *Harry Potter and the Philosopher's Stone*) arose in the late eighteenth century. Roger Bacon refers to an Arabic author named Artephius, whom he claims to have lived to the age of 1,025. See Gerald J. Gruman, *A History of Ideas about the Prolongation of Life* (Philadelphia: American Philosophical Society, 1966; reprint, New York: Arno Press, 1977), esp. pp. 28–68; Agostino Paravicini Bagliani, "Ruggero Bacone e l'alchimia di lunga vita: Riflessioni sui testi," in Crisciani and Bagliani, "Alchimia e medicina nel Medioevo," 33–54; and Pereira, "Tesoro inestimabile."

57. Pliny, *Natural History*, book 36, chapter 66. On glass and alchemy, see Beretta, *Alchemy of Glass*.

58. No historical Abbot Cremer has ever been identified; an alchemical treatise purporting to be his and that tells the Lull legend appeared in the seventeenth century: *Testamentum Cremeri*, published by Michael Maier in his *Tripus aureus* (Frankfurt, 1618), republished in *Musaeum hermeticum* (Frankfurt, 1678; reprint, Graz: Akademische Druck, 1970), pp. 533–44. For a lengthy version of the Lull legend, see Nicolas Lenglet du Fresnoy, *Histoire de la philosophie hermetique* (Paris, 1742–44), 1:144–84, 2:6–10, and 3:210–25; for an early version preserved in a Florentine manuscript, see the transcription in Michela Pereira, "La leggenda di Lullo alchimista," *Estudios lulianos* 27 (1987):145–63, on pp. 155–63; for a critical appraisal of its development, Pereira, *Alchemical Corpus*, pp. 38–49.

59. The linkage to the coin and a crusade overlaps with a rumor about George Ripley, a fifteenth-century alchemist and major popularizer of pseudo-Lullian alchemy in England. That story recounts that Ripley, who lived under Edward IV (the king who *did* mint the rose noble in 1464), sent £100,000 worth of alchemical gold

(prepared in the Tower of London) every year to the Knights of St. John at Rhodes to defend themselves against the Turks; see Elias Ashmole, ed., *Theatrum chemicum britannicum* (London, 1652), p. 458.

60. On the origins of these illustrations, see Barbara Obrist, *Les débuts de l'imagerie alchimique* (Paris: Le Sycomore, 1982). Intriguingly, an Arabic alchemical manuscript (falsely attributed to Zosimos) with allegorical images has recently surfaced, the first of its kind. It has been published in fascimile as *The Book of Pictures: Muṣḥaf aṣ-ṣuwar by Zosimos of Panopolis*, ed. Theodore Abt (Zurich: Living Human Heritage Publications, 2007); the editor's commentary is, however, grievously flawed and programmatic; for a scholarly analysis see Benjamin C. Hallum's learned essay review in *Ambix* 56 (2009): 76–88.

61. Pseudo-Arnald of Villanova, *Thesaurus thesaurorum et rosarium philosophorum*, in *Bibliotheca chemica curiosa*, 1:662–676; three others are found in *Bibliotheca chemica curiosa*, 2:87–134. On the first, see Antoine Calvet, "Étude d'un texte alchimique latin du XIVe siècle: Le *Rosarius philosophorum* attribué au medecin Arnaud de Villeneuve," *Early Science and Medicine* 11 (2006): 162–206.

62. A facsimile edition is *Rosarium philosophorum: Ein alchemisches Florilegium des Spätmittelalters*, ed. Joachim Telle, 2 vols. (Weinheim: VCH, 1992). This edition contains a German translation of the text, an excellent essay by Telle, and useful bibliographical information. Telle's essay appears in French translation as "Remarques sur le *Rosarium philosophorum* (1550)," *Chrysopoeia* 5 (1992–96): 265–320.

63. *Rosarium*, pp. 46–47.

64. Ibid., pp. 46 and 55.

65. For more on this topic, see Lawrence M. Principe, "Revealing Analogies: The Descriptive and Deceptive Roles of Sexuality and Gender in Latin Alchemy," in *Hidden Intercourse: Eros and Sexuality in the History of Western Esotericism*, ed. Wouter J. Hanegraaff and Jeffrey J. Kripal (Leiden: Brill, 2008), pp. 208–29.

66. Albert the Great, *Mineralia*, book 4, chapter 1; in *Alberti Magni opera omnia*, 5:83.

67. Ibid., 5:84.

68. Albert, *Physica*, book 1, tractate 3, chapter 12; in *Alberti Magni opera omnia*, 3:72; *Mineralia*, book 1, tractate 1, chapter 5; in ibid., 5:7. See also Obrist, *Débuts*, pp. 31–33.

69. Even Zosimos refers to a substance under the name of hermaphrodite (*arsenothēlu*); Mertens, *Les alchimistes grecs IV, i: Zosime*, p. 21. In that case, he is probably using hermaphrodite as a *Deckname* for mercury, drawing upon well-known astrological ideas that some planets are "male" (Sun, Mars, Jupiter, Saturn) and some "female" (Moon and Venus), while Mercury is common to both genders because it "produces the dry and the moist alike"; see Ptolemy, *Tetrabiblos*, 1:6. For more on this topic, see Achim Aurnhammer, "Zum Hermaphroditen in der Sinnbildkunst der Alchemisten," in *Die Alchemie in der europäischen Kultur- und Wissenschaftsgeschichte*, ed. Christoph Meinel, Wolfenbütteler Forschungen 32 (Wiesbaden: Harrassowitz, 1986), pp. 179–200, and Leah DeVun, "The Jesus Hermaphrodite: Science and Sex Difference in Premodern Europe," *Journal for the History of Ideas* 69 (2008): 193–218.

70. Wilhelm Ganzenmüller, "Das Buch der heiligen Dreifaltigkeit," *Archiv der Kulturgeschichte* 29 (1939): 93–141; Herwig Buntz, "Das *Buch der heiligen Dreifaltigkeit*, sein Autor und seine Überlieferung," *Zeitschrift für deutsches Altertums und deutsche Literatur* 101 (1972): 150–60; Marielene Putscher, "Das *Buch der heiligen Dreifaltigkeit* und seine Bilder in Handschriften des 15. Jahrhunderts," in Meinel, *Die Alchemie in*

der europäischen Kultur- und Wissenschaftsgeschichte, pp. 151–78; and Obrist, *Débuts*, pp. 117–82.

Chapter Four

1. On Libavius, see Bruce T. Moran, *Andreas Libavius and the Transformation of Alchemy: Separating Chemical Cultures with Polemical Fire* (Sagamore Beach, MA: Science History Publications, 2007).

2. William R. Newman and Lawrence M. Principe, "Alchemy vs. Chemistry: The Etymological Origins of a Historiographic Mistake," *Early Science and Medicine* 3 (1998): 32–65. See also Halleux, *Les textes alchimiques*, pp. 43–49.

3. Newman and Principe, "Etymological Origins," pp. 43–44. Part of the problem is that in the past, historians have too often assigned historical characters, books, or topics to one category or the other based on arbitrary and anachronistic presuppositions, resulting in the projection into the past of a false dichotomy based on modern ideas, and a consequential skewing of our historical understanding. Once we start talking about an inclusive *chymistry*, many apparent problems and conundrums vanish, and we can better operate within the historical context to come to more accurate understandings. For an example of how this works in the case of Isaac Newton, see Lawrence M. Principe, "Reflections on Newton's Alchemy in Light of the New Historiography of Alchemy," in *Newton and Newtonianism: New Studies*, ed. James E. Force and Sarah Hutton (Dordrecht: Kluwer, 2004), pp. 205–19.

4. Robert Boyle, "Essay on Nitre," from *Certain Physiological Essays* (1661), in *The Works of Robert Boyle*, ed. Michael Hunter and Edward B. Davis, vol. 2 (London: Pickering and Chatto, 1999), 85.

5. The very cursory account given here of the moral attacks on alchemy and their relation to the professionalization of chemistry at this time is treated more fully in Lawrence M. Principe, "A Revolution Nobody Noticed? Changes in Early Eighteenth Century Chymistry," in *New Narratives in Eighteenth-Century Chemistry*, ed. Lawrence M. Principe (Dordrecht: Springer, 2007), pp. 1–22, and at greater length in my forthcoming *Wilhelm Homberg and the Transmutations of Chymistry*. See also John C. Powers, "'Ars sine Arte': Nicholas Lemery and the End of Alchemy in Eighteenth-Century France," *Ambix* 45 (1998): 163–89.

6. Étienne-François Geoffroy, "Des supercheries concernant la pierre philosophale," *Mémoires de l'Académie Royale des Sciences* 24 (1722): 61–70.

7. Principe, *Wilhelm Homberg* (forthcoming), and until that time, "Transmuting Chymistry into Chemistry: Eighteenth-Century Chrysopoeia and Its Repudiation," in *Neighbours and Territories: The Evolving Identity of Chemistry*, ed. José Ramón Bertomeu-Sánchez, Duncan Thorburn Burns, and Brigitte Van Tiggelen (Louvain-la-Neuve, Belgium: Mémosciences, 2008), pp. 21–34.

8. James Price, *An Account of some Experiments on Mercury, Silver and Gold, made in Guildford in May, 1782* (Oxford, 1782); P. J. Hartog and E. L. Scott, "Price, James (1757/8–1783)," *Oxford Dictionary of National Biography* (Oxford: Oxford University Press, 2004), corrects some errors in the longer accounts by Denis Duveen, "James Price (1752–1783) Chemist and Alchemist," *Isis* 41 (1950): 281–83, and H. Charles Cameron, "The Last of the Alchemists," *Notes and Records of the Royal Society* 9 (1951): 109–14; the latter takes a particularly and unhelpfully cynical view of the affair.

9. On the construction of the category of occult sciences and its rejection by the academic establishment, see Wouter J. Hanegraaff, *Esotericism and the Academy:*

Rejected Knowledge in Western Culture (Cambridge: Cambridge University Press, 2012), esp. 184ff.

10. Johann Christoph Adelung, *Geschichte der menschlichen Narrheit; oder, Lebensbeschreibungen berühmter Schwarzkünstler, Goldmacher, Teufelsbanner, Zeichen- und Liniendeuter, Schwärmer, Wahrsager, und anderer philosophischer Unholden*, 7 vols. (Leipzig, 1785–89).

11. "Der Goldmacher zu London," *Teutsche Merkur*, February 1783, pp. 163–91.

12. Lawrence M. Principe, "Alchemy Restored," *Isis* 102 (2011): 305–12.

13. Johann Christian Wiegleb, *Historisch-kritische Untersuchung der Alchimie* (Weimar, 1777; reprint, Leipzig: Zentral-Antiquariat der DDR, 1965). For an analysis of this work, see Dietlinde Goltz, "Alchemie und Aufklärung: Ein Beitrag zur Naturwissenschaftsgeschichtsschreibung der Aufklärung," *Medizinhistorische Journal* 7 (1972): 31–48. Also, Achim Klosa, *Johann Christian Wiegleb (1732–1800): Ein Ergobiographie der Aufklärung* (Stuttgart: Wissenschaftliche Buchgesellschaft, 2009).

14. One example of many is *Das Geheimnis aller Geheimnisse . . . oder der güldene Begriff der geheimsten Geheimnisse der Rosen- und Gülden-Kreutzer* (Leipzig, 1788), which is a compendium of laboratory recipes and advice for chrysopoetic and medicinal arcana.

15. For studies of these groups, see Renko Geffarth, *Religion und arkane Hierarchie: Der Orden der Gold- und Rosenkreuzer als geheime Kirche im 18. Jahrhundert* (Leiden: Brill, 2007); Christopher McIntosh, *The Rose Cross and the Age of Reason: Eighteenth Century Rosicrucianism in Central Europe and Its Relationship to the Enlightenment* (Leiden: Brill, 1992); Antoine Faivre, ed., *René Le Forestier, La Franc-Maçonnerie templière et occultiste aux XVIIIᵉ et XIXᵉ siècles* (Paris: Aubier-Montaigne, 1970), also available in German translation as Alain Durocher and Antoine Faivre, eds., *Die templerische und okkultistische Freimaurerei im 18. und 19. Jahrhundert*, 4 vols. (Leimen: Kristkeitz, 1987–92); H. Möller, "Die Gold- und Rosenkreuzer, Struktur, Zielsetzung und Wirkung einer anti-aufklärerischen Geheimgesellschaft," in *Geheime Gesellschaften*, ed. Peter Christian Ludz (Heidelberg: Schneider, 1979), pp. 153–202; and Hanegraaff, *Esotericism*, pp. 211–12.

16. Andreas Ruff, *Die neuen kürzeste und nützlichste Scheide-Kunst oder Chimie theoretisch und practisch erkläret* (Nuremberg, 1788), p. 200.

17. Among the most famous connections to alchemy in this period is the study of the subject by Goethe as described in his autobiographical *Dichtung und Wahrheit*, for example vol. 1, bk. 8 and vol. 2, bk. 10. See also Rolf Christian Zimmermann, *Das Weltbild des jungen Goethe: Studien zur hermetischen Tradition des deutschen 18. Jahrhunderts*, 2 vols. (Munich: Wilhelm Fink, 1969–79). It is also noteworthy that Mary Shelley's Frankenstein begins his arcane studies by reading notable alchemical authors.

18. For example, L. P. François Cambriel, *Cours de philosophie hermétique ou d'alchimie* (Paris, 1843), and Cyliani, *Hermès dévoilé* (Paris, 1832; reprint, Paris: Éditions Traditionnelles, 1975); the latter is so self-consciously literary in character, and so freely borrows tropes from early modern alchemical literature, that even many later adherents of alchemy tended to see it as a literary production rather than a description of personal experience and practices. Louis Lucas's *La chimie nouvelle* (Paris, 1854) is sometimes cited as an "alchemical" work of the period, but it is actually an attempt to ground an entirely new philosophy of nature, largely through a new physics, and only glancingly mentions the possibility of metallic transmutation as a conse-

quence of the idea that the elements are sequential configurations of hydrogen (pp. 402–4), an idea that was already being discussed by established chemists.

19. Albert Poisson, *Thèories et symboles des alchimistes* (Paris, 1891). His date of death is variously given; see Richard Caron, "Notes sur l'histoire de l'alchimie en France à la fin du XIXᵉ et au début du XXᵉ siècle," in *Ésotérisme, gnoses & imaginaire symbolique*, ed. Richard Caron, Joscelyn Godwin, Wouter J. Hanegraaff, and Jean-Louis Vieillard-Baron (Leuven: Peeters, 2001), pp. 17–26, esp. p. 20. Georges Richet ("La science alchimique au XXᵉ siècle," in *Le voile d'Isis*, December 1922), claims Poisson died at age twenty-nine in 1894. More widely on late alchemy, see Caron's article "Alchemy V: 19th and 20th Century," in *Dictionary of Gnosis and Western Esotericism*, ed. Wouter J. Hanegraaff et al. (Leiden: Brill, 2005), 1:50–58.

20. For example, Archibald Cockren, *Alchemy Rediscovered and Restored* (London: Rider, 1940) [this text is more chemiatric than chrysopoetic, but follows early modern ideas predominantly], and Lapidus, *In Pursuit of Gold: Alchemy in Theory and Practice* (New York: Samuel Weiser, 1976).

21. Cyprien Théodore Tiffereau, *Les métaux sont des corps composés* (Vaugirard, 1855; reprinted as *L'or et la transmutation des métaux* [Paris, 1889]). He published his first memoir in 1853 as the eight-page pamphlet *Les métaux ne sont pas des corps simples*; his 1855 publication collects six memoirs presented to the Académie des Sciences; the 1889 edition includes additional materials and the transcript of a public lecture he gave in that year.

22. For example, Alexandre Baudrimont, *Traité de chimie générale et expérimentale* (Paris, 1844), 1:68–69 and 275.

23. Victor Meunier, *La Presse*, June 24, 1854; reprinted in Tiffereau, *Les métaux sont des corps composés*, p. xix.

24. C. Théodore Tiffereau, *L'art de faire l'or* (Paris, 1892), pp. 61 and 89–102. He notes that he was inspired by Edouard Trouessart's *Les microbes, les ferments, et les moisissures* (Paris, 1886) and by Pasteur's findings.

25. George B. Kauffman, "The Mystery of Stephen H. Emmens: Successful Alchemist or Ingenious Swindler?," *Ambix* 30 (1983): 65–88.

26. Louis Figuier, *L'Alchimie et les alchimistes*, 2nd ed. (Paris, 1856), pp. 343–75.

27. The examination of alchemy in Victorian occultism is covered at greater length in Lawrence M. Principe and William R. Newman, "Some Problems in the Historiography of Alchemy," in *Secrets of Nature: Astrology and Alchemy in Early Modern Europe*, ed. William Newman and Anthony Grafton (Cambridge, MA: MIT Press, 2001), pp. 385–434.

28. A fragment of Thomas South's poem was found in 1918 as proof sheets folded into a secondhand book in a London bookshop. This fragment was published by William Leslie Wilmshurst in *The Quest* 10 (1919): 213–25, and reprinted by the Alchemical Press (Edmonds, WA) in 1984.

29. Mary Anne Atwood, *A Suggestive Inquiry into the Hermetic Mystery: With a dissertation on the more celebrated of the alchemical philosophers being an attempt towards the recovery of the ancient experiment of nature* (London: T. Saunders, 1850). The first reprint (Belfast: William Tait, 1918) contains an introduction by Walter Leslie Wilmshurst, in which the above explanation of the destruction of the first printing is given on pp. 6–9. A revised edition appeared in 1920, and a reprint thereof in 1960 (New York: Julian Press). An undated reprint (of the 1918 edition) was made by the Yogi Publication Society.

30. Ibid., p. 26. This and subsequent references are to the reprint edition of 1918.

31. Ibid., pp. 78–85, 96–98, 162, 454–55.

32. Ibid., p. 162.

33. Ibid., p. 143.

34. On Mesmer and animal magnetism, see Hanegraaff et al., eds., *Dictionary of Gnosis*, 1:76–82 and references therein; also Alison Winter, *Mesmerized: Powers of Mind in Victorian Britain* (Chicago: University of Chicago Press, 1998).

35. Atwood, *Suggestive Inquiry*, p. 543. For the broader background of this attempt to read animal magnetism into historical documents, see what has been called "magnetic historiography" in Hanegraaff, *Esotericism*, pp. 260–77, esp. 266–77.

36. Atwood, *Suggestive Inquiry*, pp. 527–28.

37. Ethan Allen Hitchcock, *Remarks upon Alchymists* (Carlisle, PA, 1855); *Remarks upon Alchemy and the Alchemists* (Boston, 1857; reprint, New York: Arno Press, 1976). The quotation cited is on p. 19. The review appeared in *Westminster Review* 66 (October 1856): 153–62. See I. Bernard Cohen, "Ethan Allen Hitchcock: Soldier-Humanitarian-Scholar, Discoverer of the 'True Subject' of the Hermetic Art," *Proceedings of the American Antiquarian Society* 61 (1951): 29–136.

38. Hitchcock, *Remarks upon Alchemy*, pp. iv–v.

39. Ibid., pp. viii and 30.

40. For more detail on alchemy in Victorian occultist circles, see Principe and Newman, "Some Problems," pp. 388–401.

41. S. A. [Sapere Aude, pseudonym of William Wynn Westcott], *The Science of Alchymy* (London: Theosophical Publishing Society, 1893). Westcott was also the coroner for North East London and coeditor of *The Extra Pharmacopaeia of Unofficial Drugs* (1883); he is mentioned (with a photograph) in *The Chemist and Druggist* for September 2, 1922, p. 339.

42. Ellic Howe, *The Magicians of the Golden Dawn* (New York: Samuel Weiser, 1978); Ellic Howe, ed., *The Alchemist of the Golden Dawn: The Letters of the Reverend W. A. Ayton to F. L. Gardner and Others 1886–1905* (Wellingborough, UK: Aquarian Press, 1985); R. A. Gilbert, *The Golden Dawn: Twilight of the Magicians* (San Bernardino, CA: Borgo Press, 1988). See also the issues of *Cauda Pavonis* dedicated to the Golden Dawn: new series 8, Spring and Fall 1989 and Spring 1990.

43. See for example Christopher McIntosh, *Eliphas Lévi and the French Occult Revival* (London: Rider, 1975); specifically in terms of alchemy, see the useful overview in M. E. Warlick, *Max Ernst and Alchemy: A Magician in Search of a Myth* (Austin: University of Texas Press, 2001), pp. 21–33.

44. Hargrave Jennings, *The Rosicrucians* (London, 1870); see esp. pp. 20–39; Albert Pike, *Morals and Dogma of the Ancient and Accepted Scottish Rite* (London, 1871).

45. See R. A. Gilbert, *A. E. Waite: Magician of Many Parts* (Wellingborough, UK: Crucible, 1987).

46. Arthur Edward Waite, *Lives of the Alchemystical Philosophers* (London, 1888), pp. 9–37, 273. The work was reissued as *Alchemists through the Ages* (New York: Rudolf Steiner Publications, 1970).

47. Ibid., pp. 30–37 and 273–75; A. E. Waite, *Azoth; or, The Star in the East, Embracing the First Matter of the Magnum Opus, the Evolution of the Aphrodite-Urania, the Supernatural Generation of the Son of the Sun, and the Alchemical Transfiguration of Humanity* (London, 1893; reprint, Secaucus, NJ: University Books, 1973), pp. 54, 58, and 60.

48. A. E. Waite, *The Secret Tradition of Alchemy* (New York: Alfred Knopf, 1926), p. 366.

49. Waite's oral remarks about papers delivered at meetings of the Alchemical Society from 1913 until 1915 (and recorded in its *Journal*) indicate a considerably more critical and historically sophisticated approach than is evident in his nineteenth-century publications.

50. For a brief account of some such writers, see Halleux, *Les textes alchimiques*, pp. 56–58.

51. The journal was published under several names at different times: *L'Hyperchimie, Les nouveaux horizons de la science et de la pensée, Rosa alchemica*, and after 1920, *La Rose+Croix*.

52. François Jollivet-Castelot, *Comment on devient alchimiste* (Paris, 1897); *La synthèse de l'or* (Paris: Daragon, 1909); *La révolution chimique et la transmutation des métaux* (Paris: Chacornac, 1925), this volume contains "La philosophie alchimique" (pp. 46–52), which gives a concise summary of his views on alchemy, and 175–78 describes his Societé; and *Synthèse des sciences occultes* (Paris, 1928). For an analysis, see Richard Caron, "Notes," pp. 23–26.

53. The famous historian of alchemy John Ferguson was the president, and the honorary vice presidents included A. E. Waite and Isabelle de Steiger, a former associate of Mary Anne Atwood. The *Journal of the Alchemical Society* was published in twenty-one issues; the first appeared in January 1913 and the last (a double issue) in September 1915.

54. On the fascinating linkage of early twentieth-century chemistry and physics with Victorian occultism, including within the Alchemical Society of London, see the excellent study by Mark S. Morrisson, *Modern Alchemy: Occultism and the Emergence of Atomic Theory* (Oxford: Oxford University Press, 2007). For a contemporaneous overview of radioactivity and alchemy, see Jollivet-Castelot, *Révolution*, "Les thèories modernes de l'alchimie," pp. 179–98. On radium's comparison to the Philosophers' Stone, see for example Fritz Paneth, "Ancient and Modern Alchemy," *Science* 64 (1926): 409–17, esp. 415.

55. Carl Gustav Jung, "Die Erlösungsvorstellungen in der Alchemie," *Eranos-Jahrbuch 1936* (Zurich: Rhein-Verlag, 1937), pp. 13–111, quoting from p. 17. In the later English version, "The Idea of Redemption in Alchemy," in *The Integration of the Personality*, ed. Stanley Dell (New York: Farrar and Rinehart, 1939), pp. 205–80, quoting from p. 210, the claim is made more strongly, probably under Jung's direction. All of Jung's contributions on alchemy are found in *The Collected Works of Carl Gustav Jung* (London: Routledge, 1953–79), vol. 9, pt. 2: *Aion*; vol. 12: *Psychology and Alchemy*; vol. 13: *Alchemical Studies*; vol. 14: *Mysterium Conjunctionis*. For further analysis of Jung's views on alchemy, see Principe and Newman, "Some Problems," pp. 401–8.

56. Jung, "Erlösungsvorstellungen," pp. 19, 20, 23–24; "Idea of Redemption," pp. 212, 213, 215.

57. Jung, "Erlösungsvorstellungen," p. 20; "Idea of Redemption," pp. 212–13.

58. Jung, "Erlösungsvorstellungen," p. 60; "Idea of Redemption," p. 239.

59. Luther H. Martin, "A History of the Psychological Interpretation of Alchemy," *Ambix* 22 (1975): 10–20, esp. 12–16; F. X. Charet, *Spiritualism and the Foundations of C. G. Jung's Psychology* (Albany, NY: SUNY Press, 1993); Herbert Silberer, *Hidden Symbolism of Alchemy and the Occult Arts* (New York: Dover, 1971; originally published 1917 as *Problems of Mysticism and Its Symbolism*); Richard Noll, *The Jung Cult* (Princeton, NJ:

Princeton University Press, 1994), pp. 144 and 171; *The Aryan Christ* (New York: Random House, 1997), pp. 25–30, 37–41, 229–30.

60. Mircea Eliade, "Metallurgy, Magic and Alchemy," *Cahiers de Zalmoxis*, 1 (Paris: Librairie Orientaliste Paul Geuthner, 1938), quoting from p. 44. This early work was elaborated into the widely popular *The Forge and the Crucible* (Chicago: University of Chicago Press, 1978 [first English publication, 1962]), quoting from p. 162, originally published as *Forgerons et alchimistes* (Paris: Flammarion, 1956); for the occultist roots, see Mac Linscott Ricketts, *Mircea Eliade: The Romanian Roots, 1907–1945* (Boulder, CO: East European Monographs, 1988), pp. 141–53, 313–25, 804–8, 835–42; for the explicit influences of Jung, see for example *Forge and Crucible*, pp. 52, 158, 161, 163, and 221–26.

61. For more on Eliade, see Principe and Newman, "Some Problems," pp. 408–15, and Obrist, *Débuts*, pp. 12–33.

62. Israel Regardie, *The Philosopher's Stone: A Modern Comparative Approach to Alchemy from the Psychological and Magical Points of View* (London: Rider, 1938).

63. Ibid., pp. 18–19.

64. Morrisson, *Modern Alchemy*, pp. 188–91.

65. See for example Obrist, *Débuts*, esp. pp. 11–21 and 33–36; Principe and Newman, "Some Problems," pp. 401–8; Dan Merkur, "Methodology and the Study of Western Spiritual Alchemy," *Theosophical History* 8 (2000): 53–70; Halleux, *Les textes alchimiques*, pp. 55–58; Harold Jantz, "Goethe, Faust, Alchemy, and Jung," *German Quarterly* 35 (1962): 129–41.

66. Besides the influence of the Golden Dawn on the poet Yeats, mentioned above, the alchemy of Jollivet-Castelot attracted and influenced Swedish playwright August Strindberg (their correspondence is published as August Strindberg, *Bréviaire alchimique*, ed. François Jollivet-Castelot [Paris: Durville, 1912], and see Alain Mercier, "August Strindberg et les alchimistes français: Hemel, Vial, Tiffereau, Jollivet-Castelot," *Revue de littérature comparée* 43 [1969]: 23–46), and occultist alchemy forms a background to the work of the artist Max Ernst; see Warlick, *Max Ernst*.

67. On this third revival, see Bruce T. Moran, "Alchemy and the History of Science: Introduction," *Isis* 102 (2011): 300–304; Principe, "Alchemy Restored," ibid., 305–312; and Marcos Martinón-Torres, "Some Recent Developments in the Historiography of Alchemy," *Ambix* 58 (2011): 215–37.

Chapter Five

1. For a quick introduction to the history of science in the Scientific Revolution, see Lawrence M. Principe, *The Scientific Revolution: A Very Short Introduction* (Oxford: Oxford University Press, 2011); for more detail, see Margaret J. Osler, *Reconfiguring the World: Nature, God, and Human Understanding from the Middle Ages to Early Modern Europe* (Baltimore: Johns Hopkins University Press, 2010).

2. For a wide-ranging overview of the breadth of ideas, characters, and activities encompassed by early modern alchemy, see Bruce T. Moran, *Distilling Knowledge: Alchemy, Chemistry, and the Scientific Revolution* (Cambridge, MA: Harvard University Press, 2005).

3. On alchemy's exile from and return to the fold of the history of science, see Principe, "Alchemy Restored."

4. In fact, by the end of the seventeenth century, chymists had also discovered zinc, bismuth, and possibly cobalt; but these were not grouped with the classical seven

metals, and were sometimes called "bastard metals" because they did not fully share the characteristic properties of shine and malleability with the others.

5. Vladimir Karpenko, "Systems of Metals in Alchemy," *Ambix* 50 (2003): 208–30.

6. The earliest mention I have found of this idea is by the historian ibn-Khaldūn, writing in 1367 (but he must have been borrowing from an earlier text), who says that gold takes 1,080 years to form, corresponding with the period of a particular solar cycle; *The Muqaddimah*, 3:274. The connection between planets and the formation of the corresponding metals is exaggerated and ridiculed by Nicolas Lemery, *Cours de chymie* (Paris, 1683), pp. 69–71.

7. Alain-Philippe Segonds, "Astronomie terrestre/Astronomie céleste chez Tycho Brahe," in *Nouveau ciel, nouvelle terre: La révolution copernicienne dans l'allemagne de la réforme (1530–1630)*, ed. Miguel Ángel Granada and Édouard Mehl (Paris: Les Belles Lettres, 2009, pp. 109–42; "Tycho Brahe et l'alchimie," in Margolin and Matton, *Alchimie et philosophie à la Renaissance*, pp. 365–78. On Tycho's laboratory, see Jole Shackelford, "Tycho Brahe, Laboratory Design, and the Aim of Science: Reading Plans in Context," *Isis* 84 (1993): 211–30.

8. For a warning against *particularia*, Gaston Duclo, *De triplici praeparatione argenti et auri*, in *Theatrum chemicum*, 4:371–388, esp. 374–75; for the balance sheets, see *Coelum philosophorum* (Frankfurt and Leipzig, 1739), pp. 60, 125–26.

9. Published in Principe, *The Aspiring Adept*, pp. 302–4 (orthography slightly modernized here). For more on *particularia*, see ibid., pp. 77–80. The main text of Boyle's list of particulars is lost, possibly pilfered by some aspiring transmuter of metals who had access to his papers; today only the prefatory material remains. See Michael Hunter and Lawrence M. Principe, "The Lost Papers of Robert Boyle," *Annals of Science* 60 (2003): 269–311.

10. On potable gold, see Angelo Sala, *Processus de auro potabili* (Strasbourg, 1630); *De auro potabili* in *Theatrum chemicum*, 6:382–93; Francis Anthony, *The apologie, or defence of . . . aurum potabile* (London, 1616); Guglielmo Fabri, *Liber de lapide philosophorum et de auro potabili*, in *Il Papa e l'alchimia: Felice V, Guglielmo Fabri e l'elixir*, by Chiara Crisciani (Rome: Viella, 2002), pp. 118–83, citing pp. 150–60 [Latin text and Italian translation of a mid-fifteenth-century text on alchemy dedicated to anti-Pope Felix V]; and Ernst Darmstaedter, "Zur Geschichte des *Aurum potabile*," *Chemiker-Zeitung* 48 (1924): 653–55, 678–80.

11. Several accounts of this extraction are given by the polymath Daniel Georg Morhof in his brief but rich study of chrysopoeia originally published in 1671, *De metallorum transmutatione*, in *Bibliotheca chemica curiosa*, 1:168–92, esp. p. 178. A private account of the removal of the tincture from a gold coin, leaving the coin white, is recounted in the February 26, 1680, letter of one De Saintgermain to Robert Boyle, published in *The Correspondence of Robert Boyle*, ed. Michael Hunter, Lawrence M. Principe, and Antonio Clericuzio (London: Pickering and Chatto, 2001), 5:185–90. See also Principe, *Aspiring Adept*, pp. 82–86.

12. Joan Baptista Van Helmont, *Opuscula medica inaudita* (Amsterdam, 1648; reprint, Brussels: Culture et Civilization, 1966), "De lithaisi," pp. 69ff.; Principe, *Aspiring Adept*, pp. 88–89.

13. On Starkey, see William R. Newman, *Gehennical Fire: The Lives of George Starkey, an American Alchemist in the Scientific Revolution* (Cambridge, MA: Harvard University Press, 1994), and Newman and Lawrence M. Principe, *Alchemy Tried in the Fire: Starkey, Boyle, and the Fate of Helmontian Chymistry* (Chicago: University of

Chicago Press, 2002); on *luna fixa*, see George Starkey, *Alchemical Laboratory Notebooks and Correspondence*, ed. William R. Newman and Lawrence M. Principe (Chicago: University of Chicago Press, 2004), pp. xxiii–xxxiv, and Morhof, *De metallorum transmutatione*, 1:187.

14. One well-documented example is an attempt in 1684 by Gottfried von Sonnenberg to sell a recipe for the Philosophers' Stone for £7,000 to Robert Boyle or another member of the Royal Society of London; see Principe, *Aspiring Adept*, pp. 114–15, and Boyle, *Correspondence*, 6:52–86 and 116–21.

15. One citation of this phrase appears in Thomas Norton, *Ordinall of Alchimy*, in Ashmole, *Theatrum chemicum britannicum*, pp. 1–106, on p. 87, although attributed there to "Maria Sister of Aron."

16. For an early modern example involving the analysis of alchemical laboratory notebooks, see Newman and Principe, *Alchemy Tried in the Fire*, pp. 100–155.

17. Pseudo-Arnald of Villanova, *De secretis naturae*, ed. and trans. Antoine Calvet, in "Cinq traités alchimique médiévaux," *Chrysopoeia* 6 (1997–99): 154–206; "dicam ut fateos derideam, sapientes doceam," p. 178.

18. Richard S. Westfall, "Alchemy in Newton's Library," *Ambix* 31 (1994): 97–101; on Newton's alchemy more generally, see Betty Jo Teeter Dobbs, *The Foundations of Newton's Alchemy; or, Hunting of the Greene Lyon* (Cambridge: Cambridge University Press, 1975) and *The Janus Faces of Genius* (Cambridge: Cambridge University Press, 1991). Many of Dobbs's conclusions have had to be revised in the light of more recent studies; see for example Principe, "Reflections on Newton's Alchemy," and William R. Newman, "Newton's *Clavis* as Starkey's *Key*." *Isis* 78 (1987): 564–74.

19. Classifying chrysopoeians on the basis of their favored starting material was done notably by Georg Ernst Stahl in his *Fundamenta chymiae dogmaticae* (Leipzig, 1723), translated by Peter Shaw as *Philosophical Principles of Universal Chemistry* (London, 1730); see Kevin Chang, "The Great Philosophical Work: Georg Ernst Stahl's Early Alchemical Teaching," in López-Pérez, Kahn, and Rey Bueno, *Chymia*, pp. 386–96.

20. Abbot Cremer, *Testamentum Cremeri*, pp. 531–44, quoting from p. 535.

21. Pseudo-Nicolas Flamel, *Exposition of the Hieroglyphicall Figures* (London, 1624; reprint, New York: Garland, 1994), pp. 11–13. On Flamel and his legend, see Robert Halleux, "Le mythe de Nicolas Flamel, ou les méchanismes de la pseudépigraphie alchimique," *Archives internationales de l'histoire des sciences* 33 (1983): 234–55.

22. On this remarkable tale, see Principe, *Aspiring Adept*, pp. 115–34; "Georges Pierre des Clozets, Robert Boyle, the Alchemical Patriarch of Antioch, and the Reunion of Christendom: Further New Sources," *Early Science and Medicine* 9 (2004): 307–20; and Noel Malcolm, "Robert Boyle, Georges Pierre des Clozets, and the Asterism: New Sources," ibid., 293–306.

23. On Dee and angels, see Deborah Harkness, *John Dee's Conversations with Angels: Cabala, Alchemy, and the End of Nature* (Cambridge: Cambridge University Press, 1999); on Boyle and angels, Principe, *Aspiring Adept*, pp. 195–97 and 310–17 (Boyle's "Dialogue on the Converse with Angels"); and Michael Hunter, "Alchemy, Magic, and Moralism in the Thought of Robert Boyle," *British Journal for the History of Science* 23 (1990): 387–410.

24. Athanasius Kircher, *Mundus subterraneus* (Amsterdam, 1678), pp. 301–2.

25. Martin Del Rio, *Disquisitionum magicarum libri sex* (Ursel, 1606), bk. 1, chap. 5; *Investigations Into Magic*, trans. and ed. P. G. Maxwell-Stuart (Manchester, NY:

Manchester University Press, 2000), this useful book is not a complete translation but rather a series of translated passages linked by condensed summaries of the intervening texts; Martha Baldwin, "Alchemy and the Society of Jesus in the Seventeenth Century: Strange Bedfellows?," *Ambix* 40 (1993): 41–64. For a valuable compendium of texts by Jesuits both for and against transmutational alchemy, see Sylvain Matton, *Scolastique et alchimie*, Textes et Travaux de Chrysopoeia 10 (Paris: SÉHA; Milan: Archè, 2009), esp. pp. 1–76.

26. Nicolas Eymerich, *Contra alchemistas*, ed. Sylvain Matton, *Chrysopoeia* 1 (1987): 93–136, quoting from pp. 132–33 [Latin text with facing-page French translation]. For a fascinating view of the relationship of demonic power with alchemy, see Newman, *Promethean Ambitions*, pp. 47–62 and 91–97.

27. Newman, *Gehennical Fire*, pp. 87–90 and 212–26. For the latest biographical information on Sendivogius, see the work of Rafal T. Prinke—for example, "Beyond Patronage: Michael Sendivogius and the Meanings of Success in Alchemy," in López-Perez, Kahn, and Rey Bueno, *Chymia*, pp. 175–231 and references therein.

28. For medieval approaches to the Philosophers' Stone, see Michela Pereira, "Teorie dell'elixir"; one listing (and critique) of various starting materials is found in Lorenzo Ventura, *De ratione conficiendi lapidis philosophici*, in *Theatrum chemicum*, 2:215–312, on pp. 233–39. Indeed, many chrysopoetic treatises begin with an analysis of various starting points.

29. The original source for this idea is in Morienus, *De compositione alchemiae*, 1:515; Morienus, *A Testament of Alchemy*, pp. 24–27.

30. John of Rupescissa, *De confectione*, 2:80–83, quoting from p. 80.

31. On the origins of this motto and its variations, see Joachim Telle, "Paracelsistische Sinnbildkunst: Bemerkungen zu einer Pseudo-*Tabula smaragdina* des 16. Jahrhunderts," in *Bausteine zur Medizingeschichte* (Wiesbaden: Steiner Verlag, 1984), pp. 129–39; available also in French: "L'art symbolique paracelsien: Remarques concernant une pseudo-*Tabula smaragdina* du XVIᵉ siècle," in *Présence de Hèrmes Trismégeste*, ed. Antoine Faivre (Paris: Albin Michel, 1988), pp. 184–208; Didier Kahn, "Les débuts de Gérard Dorn," in *Analecta Paracelsica: Studien zum Nachleben Theophrast von Hohenheims im deutschen Kulturgebiet der frühen Neuzeit*, ed. Joachim Telle (Stuttgart: Steiner Verlag, 1994), pp. 75–76, and "Alchemical Poetry in Medieval and Early Modern Europe: A Preliminary Survey and Synthesis; Part I: Preliminary Survey," *Ambix* 57 (2010): 263.

32. For example, in the early work *Vom grossen Stein der Uhralten* (in *Chymische Schrifften*, 1:94–98 [Hamburg, 1677; reprint, Hildesheim: Gerstenberg Verlag, 1976]; this section appeared first in 1602) *vitriol* is described with properties that only antimony displays, and so is certainly used as a *Deckname*; but in the later *Offenbahrung der verborgenen Handgriffe* (in *Chymische Schrifften*, 2:319–40), *vitriol* clearly means a metallic sulfate.

33. Jennifer Rampling, "Alchemy and 'Practical Exegesis' in Early Modern England," *Osiris* 29 (2014).

34. Nevertheless, clever alchemists managed to describe "offspring" arising even from the unlikely combination of three parents, all of them male. See Principe, "Revealing Analogies," pp. 211–14.

35. George Ripley, *Compound of Alchymie*, in Ashmole, *Theatrum Chemicum Britannicum*, pp. 107–93, quoting from pp. 130–31.

36. Ibid., p. 124.

37. *Rosarium*, 1:59.

38. The various color stages, or "regimens," of the Philosophers' Stone are particularly concisely described in the anonymous late seventeenth-century work *Stone of the Philosophers*, printed in *Collectanea chymica* (London, 1893), pp. 55–120, on pp. 113–20 and in rather more prolix fashion in Eirenaeus Philalethes [George Starkey], *Secrets Reveal'd; or, An Open Entrance to the Shut-Palace of the King* (London, 1669), pp. 80–117.

39. Elias Ashmole, annotations, *Theatrum Chemicum Britannicum*, p. 481.

40. Among the critics who thought the stone's action was unnatural (perhaps orchestrated by demons) were Meric Casaubon, *A True and Faithfull Relation* (London, 1659), preface, p. xxxx, and several Jesuits, such as Athanasius Kircher, although views among the Jesuits varied widely; see Baldwin, "Alchemy and the Society of Jesus," and Margaret Garber, "Transitioning from Transubstantiation to Transmutation: Catholic Anxieties over Chymical Matter Theory at the University of Prague," in *Chymists and Chymistry*, ed. Lawrence M. Principe (Sagamore Beach, MA: Chemical Heritage Foundation and Science History Publications, 2007), pp. 63–76.

41. One example of plusquamperfection appears in the *Rosarium*, in *Bibliotheca chemica curiosa*, 1:662–76, on p. 665.

42. For a comprehensive account of *semina*, see Hiro Hirai, *Le concept de semence dans les théories de la matière à la Renaissance de Marsile Ficin à Pierre Gassendi* (Turnhout, Belgium: Brepols, 2005).

43. Boyle, *Dialogue on Transmutation*, pp. 254–55.

44. The Paracelsus literature is vast and growing, and only a few starting points can be mentioned here. For the classic studies, see Walter Pagel, *Paracelsus: An Introduction to Philosophical Medicine in the Era of the Renaissance* (Basel: Karger, 1958), and Allen G. Debus, *The Chemical Philosophy: Paracelsian Science and Medicine in the Sixteenth and Seventeenth Centuries*, 2 vols. (New York: Science History Publications, 1977). More recent contributions include these volumes of essays: Joachim Telle, ed., *Analecta Paracelsica: Studien zum Nachleben Theophrast von Hohenheims im deutschen Kulturgebiet der frühen Neuzeit* (Stuttgart: Franz Steiner Verlag, 1994); Ole Peter Grell, ed., *Paracelsus: The Man and His Reputation, His Ideas and Their Transformation* (Leiden: Brill, 1998); Heinz Schott and Ilana Zinguer, eds., *Paracelsus und seine internationale Rezeption in der frühen Neuzeit* (Leiden: Brill, 1998), as well as Didier Kahn, *Alchimie et Paracelsisme en France (1567–1625)* (Geneva: Droz, 2007); Charles Webster, *Paracelsus: Medicine, Magic, and Mission at the End of Time* (New Haven, CT: Yale University Press, 2008); and Udo Benzenhöfer, *Paracelsus* (Reinbek: Rowohlt, 1997).

45. See for example Basil Valentine, *Vom grossen Stein der Uhralten*, in *Chymische Schrifften*, 1:12–14.

46. On this process in France, see in particular Kahn, *Alchimie et Paracelsisme*.

47. See Stephen Pumphrey, "The Spagyric Art; or, The Impossible Work of Separating Pure from Impure Paracelsianism: A Historiographical Analysis," in Grell, *Paracelsus*, pp. 21–51, and Andrew Cunningham, "Paracelsus Fat and Thin: Thoughts on Reputations and Realities," in ibid., pp. 53–77.

48. See in particular Moran, *Andreas Libavius*, esp. pp. 291–302.

49. Allen G. Debus, *The French Paracelsians* (Cambridge: Cambridge University Press, 1991), pp. 21–30. A very useful list of sources is found in Hermann Fischer, *Metaphysische, experimentelle und utilitaristische Traditionen in der Antimonliteratur zur Zeit der "wissenschaftlichen Revolution": Eine kommentierte Auswahl-Bibliographie,*

Braunschweiger Veröffenlichungen zu Geschichte der Pharmazie und der Naturwissenschaften (Brunswick, 1988).

50. Moran, *Andreas Libavius.*

51. Paracelsus[?], *De rerum natura*, in *Sämtliche Werke*, ed. Karl Sudoff; *Abteilung 1: Medizinische, wissenschaftliche, und philosophische Schriften* (Munich: Oldenbourg, 1922–33), 11:316–17.

52. See for example St. Augustine, *The City of God*, bk. 16, chap. 7; for naturalistic accounts of creation where God creates matter in the first instant, which then goes on to produce the cosmos, including life on Earth, spontaneously, see for example the authors at the twelfth-century cathedral school of Chartres, such as the *Hexaemeron* of Thierry of Chartres.

53. On the homunculus, see William R. Newman, "The Homunculus and His Forebears: Wonders of Art and Nature," in *Natural Particulars: Nature and the Disciplines in Renaissance Europe*, ed. Anthony Grafton and Nancy Siraisi (Cambridge, MA: MIT Press, 1999), pp. 321–45, and at greater length in *Promethean Ambitions*, pp. 164–237.

54. Paracelsus[?], *De rerum natura*, 11:348–49.

55. Joseph Duchesne, *Ad veritatem Hermeticae medicinae* (Paris, 1604), pp. 294–301. An earlier claim about the image of plants seen in frozen *lixivia* appears in his *Grand miroir du monde* in 1593; the relevant text is reprinted in *Ad veritatem*, p. 297.

56. Ibid., pp. 292–94.

57. For treatments of palingenesis, see Joachim Telle, "Chymische Pflanzen in der deutschen Literatur," *Medizinhistorisches Journal* 8 (1973):1–34, and Jacques Marx, "Alchimie et Palingénésie," *Isis* 62 (1971): 274–89. For a lengthy early modern summary, see Georg Franck de Franckenau and Johann Christian Nehring, *De Palingenesia* (Halle, 1717).

58. Kircher, *Mundus subterraneus*, 2:434–38.

59. Kenelm Digby, *A Discourse on the Vegetation of Plants* (London, 1661); a slightly altered recipe is given in *A Choice Collection of Rare Chymical Secrets* (London, 1682), pp. 131–32. On Digby's chymistry, see Betty Jo Teeter Dobbs, "Studies in the Natural Philosophy of Sir Kenelm Digby: Part I," *Ambix* 18 (1971): 1–25; "Part II," ibid. 20 (1973): 143–63; and "Part III," ibid. 21 (1974): 1–28.

60. Joan Baptista Van Helmont, "Pharmacopolium ac dispensatorium modernum," in *Ortus medicinae* (Amsterdam, 1648; reprint, Brussels: Culture et Civilisation, 1966), no. 13, p. 459.

61. On the willow tree, Van Helmont, "Complexionum atque mistionum elementalium figmentum," in *Ortus*, no. 30, p. 109, and Robert Halleux, "Theory and Experiment in the Early Writings of Johan Baptist Van Helmont," in *Theory and Experiment*, ed. Diderik Batens (Dordrecht: Rediel, 1988), pp. 93–101. On Helmontian experiments generally, see Newman and Principe, *Alchemy Tried in the Fire*, pp. 56–91. For a quick overview of Van Helmont, see Lawrence M. Principe, "Van Helmont," in *Dictionary of Medical Biography*, ed. W. F. Bynum and Helen Bynum (Westport, CT: Greenwood Press, 2006), 3:626–28; in more detail in Walter Pagel, *Joan Baptista Van Helmont* (Cambridge: Cambridge University Press, 1982). On seeds, see Hirai, *Le concept de semence*.

62. Van Helmont, *Opuscula*, "De lithiasi," chap. 3, no. 1, p. 20.

63. On the alkahest, see Bernard Joly, "L'alkahest, dissolvant universel, ou quand la thèorie rend pensible une pratique impossible," *Revue d'histoire des sciences* 49 (1996): 308–30; Paulo Alves Porto, "'Summus atque felicissimus salium': The Medical

Relevance of the Liquor Alkahest," *Bulletin of the History of Medicine* 76 (2002): 1–29; Principe, *Aspiring Adept*, pp. 183–84; and Newman, *Gehennical Fire*, pp. 146–48 and 181–88. For contemporaneous treatments, see George Starkey, *Liquor Alkahest* (London, 1675); Otto Tachenius, *Epistola de famoso liquore alcahest* (Venice, 1652); Jean Le Pelletier, *L'Alkaest; ou, Le dissolvant universel de Van Helmont* (Rouen, 1706); and Herman Boerhaave, *Elementa chemiae* (Paris, 1733), 1:451–61.

64. Otto Tachenius, *Hippocrates chymicus* (London, 1677), preface.

Chapter Six

1. Olaus Borrichius, *Conspectus scriptorum chemicorum celebriorum*, in *Bibliotheca chemica curiosa*, 1:38–53, esp. p. 47.

2. Georg Wolfgang Wedel, "Programma vom Basilio Valentino," in *Deutsches Theatrum Chemicum*, 1: 669–80, esp. pp. 675–76.

3. Claus Priesner, "Johann Thoelde und die Schriften des Basilius Valentinus," in Meinel, *Die Alchemie in der europäischen Kultur- und Wissenschaftsgeschichte*, pp. 107–18; Hans Gerhard Lenz, "Studien zur Lebensgeschichte des Basilius-Herausgebers Johann Thölde," in *Triumphwagen des Antimons: Basilius Valentinus, Kerckring, Kirchweger; Text, Kommentare, Studien*, ed. Lenz (Elberfeld, Germany: Humberg, 2004), pp. 272–338; and Oliver Humberg, "Neues Licht auf die Lebensgeschichte des Johann Thölde," in ibid., pp. 353–74.

4. It should be noted that in early modern nomenclature, the word *antimony* referred not to the element known by that name today but rather to its chief ore, stibnite, which is chemically antimony trisulfide.

5. The story appears commonly in nineteenth-century chemical textbooks; for example, see Robert Kane, *Elements of Chemistry* (New York, 1842), p. 384.

6. The combination of the similarity to Paracelsian ideas and the claim that Valentine lived in the fifteenth century, that is, before Paracelsus, ignited a long-running priority dispute in which Paracelsus was criticized as a plagiarist of Basil Valentine. See for example Van Helmont, *Ortus medicinae*, p. 399.

7. Valentine's recipe can be found in *Triumph-Wagen antimonii*, in *Chymische Schrifften*, 1:365–71.

8. The following account of replicating Valentine's process is covered more fully in Principe, "Chemical Translation and the Role of Impurities in Alchemy."

9. Valentine, *Triumph-Wagen antimonii*, in *Chymische Schrifften*, 1:367.

10. There is another "glass of antimony" besides the one described in early modern texts. It is ruby red, contains a much larger proportion of sulfur, and is described in chemical reference works dating after the mid-nineteenth century as "the" *vitrum antimonii* of the alchemists; see J. W. Mellor, *A Comprehensive Treatise on Inorganic and Theoretical Chemistry* (London: Longmans, 1922–37), 9:477 and *Gmelins Handbuch der anorganischen Chemie* (Leipzig: Verlag Chemie, 1924), 18B:540. All trace of the golden glass, the true glass of antimony of the early modern period, had vanished from current chemical knowledge before this investigation. It is too easy to forget that scientific knowledge does not simply accumulate from generation to generation; some is invariably forgotten, fragmented, misremembered, or lost.

11. Ores that did not contain the requisite amount of silica may still have worked for early modern chymists, because they used clay crucibles rather than modern porcelain ones—the needed silica could be dissolved out of minerals in the clay.

12. Valentine, *Triumph-Wagen antimonii*, in *Chymische Schrifften*, 1:371.

13. Basil Valentine, *Ein kurtz summarischer Tractat . . . von dem grossen Stein der Uralten* (Eisleben, 1599); there were numerous subsequent editions and translations.

14. Valentine, *Von dem grossen Stein der uhralten Weisen*, in *Chymische Schrifften*, 1:1–112, quoting from p. 24.

15. Ripley has remained, until recently, an understudied figure; for the best work, see Jennifer Rampling, "Establishing the Canon: George Ripley and His Alchemical Sources," *Ambix* 55 (2008): 189–208, and "The Catalogue of the Ripley Corpus: Alchemical Writings Attributed to George Ripley," *Ambix* 57 (2010): 125–201. The best edition of George Ripley's *Compound of Alchymie* currently available appears in Ashmole, *Theatrum Chemicum Britannicum*, pp. 107–93. It was also printed by Ralph Rabbards in 1591, a reedition of which, edited by Stanton J. Linden (Burlington, VT: Ashgate, 2001), is available.

16. Leipzig, 1602. It had been issued a second time without illustrations in *Aureum Vellus . . . Tractatus III* ([Rorschach am Bodensee; i.e. Leipzig?], 1600), pp. 610–701, and then again in Joachim Tanckius, *Promptuarium alchemiae* (Leipzig, 1610 and 1614; reprint, Graz: Akademische Druck, 1976), 2:610–702. Lambsprinck's *De lapide philosophico*, in *Musaeum hermeticum*, pp. 337–71, a product of the late sixteenth century, has a similar "stepwise" format: fifteen short chapters, each composed of an emblematic illustration and a cryptic text in verse.

17. Basil Valentine, *Von dem grossen Stein der uhralten Weisen*, in *Chymische Schrifften*, 1:7–112, quoting from p. 26.

18. Roman naturalist Pliny wrote that antimony ore was easily converted into lead (*Historia naturalis*, bk. 33, sect. 34).

19. Valentine, *Von dem grossen Stein*, 1:30–32.

20. Ibid., 1:96.

21. Ibid., 1:32.

22. Ibid., 1:34.

23. Ibid., 1:34–35; a crucial line is missing from the 1677 edition and an erroneous word inserted. For the correct text, see the 1599 edition, folio Fv; is there no end of difficulties with chrysopoetic texts?

24. John Gay, Alexander Pope, and John Arbuthnot, *Three Hours After Marriage*, ed. John Harrington Smith, Augustan Reprint Society, no. 91–92 (Los Angeles: Clark Memorial Library, 1961), p. 171.

25. Thomas Kirke Rose, "The Dissociation of Chloride of Gold," *Journal of the Chemical Society* 67 (1895): 881–904.

26. Boyle mentions "the *Aqua pugilum* [water of the fighters] aenigmatically describ'd by *Basilius*" and its ability to "elevate" gold in his *Origine of Formes and Qualities* (1666), in *Works of Robert Boyle*, 5:424.

27. The presence of ammonium salts in the "water of the fighters" further helps the gold salt to sublime successfully.

28. Besides Boyle, another successful interpreter of *Von dem grossen Stein* revealed his cleverness by publishing a highly practical exposition of the process while masquerading as Valentine himself! See *Offenbahrung der verborgenen Handgriffe* (originally published 1624), in *Chymische Schrifften*, 2:319–38. This same or another "pseudo-Valentine" produced the last work to join the Valentine corpus, the *Letztes Testament* (first published 1626), which also contains a exposition of the *Keys*.

29. Basil Valentine, *Practica cum duodecim clavibus* in Maier, *Tripus aureus*, pp. 7–76; reprinted in *Musaeum hermeticum*, pp. 377–432.

30. Valentine, *Von dem grossen Stein*, 1:72.

31. A later reader, perhaps recognizing the crucial place of this impressive achievement in the process, chose its emblem (the rooster and fox) over all others for the place of honor in the background to a "portrait" of Basil Valentine prepared later in the seventeenth century and shown in figure 6.1.

32. For more on Starkey's life and ideas, see Newman, *Gehennical Fire*.

33. These materials are now edited, translated, and annotated in Starkey, *Alchemical Laboratory Notebooks and Correspondence*; a close analysis of some of their contents along with further information about Starkey's scientific interactions (including those with Boyle) and context in London appears in Newman and Principe, *Alchemy Tried in the Fire*.

34. Newman, *Gehennical Fire*, pp. 228–39; for a study of the larger role of chymistry in the development of particulate matter theories, see Newman, *Atoms and Alchemy*.

35. Starkey, *Alchemical Laboratory Notebooks and Correspondence*, pp. 228–60; Newman and Principe, *Alchemy Tried in the Fire*, pp. 188–97.

36. On the mercuralist school, see Georg Ernst Stahl, *Philosophical Principles of Universal Chemistry*, trans. Peter Shaw (London, 1730), pp. 401–16, and Lawrence M. Principe, "Diversity in Alchemy: The Case of Gaston 'Claveus' DuClo, a Scholastic Mercurialist Chrysopoeian," in *Reading the Book of Nature: The Other Side of the Scientific Revolution*, ed. Allen G. Debus and Michael Walton (Kirksville, MO: Sixteenth Century Press, 1998), pp. 181–200.

37. Principe, *Aspiring Adept*, pp. 153–55.

38. Eirenaeus Philalethes [George Starkey], *Introitus apertus ad occlusum regis palatium*, in *Musaeum hermeticum*, pp. 647–99, quoting from pp. 658–59.

39. Starkey, *Alchemical Laboratory Notebooks and Correspondence*, pp. 12–31, quoting from pp. 22–23; Boyle, *Correspondence*, 1:90–103.

40. Principe, *Aspiring Adept*, pp. 158–79; Newman, "Newton's *Clavis* as Starkey's Key."

41. Ripley, *Compound of Alchymie*, p. 141.

42. Jean Collesson, *Idea perfecta philosophiae hermeticae*, in *Theatrum chemicum*, 6:143–62, quoting from pp. 146 and 149.

43. Eirenaeus Philalethes [George Starkey], *Ripley Reviv'd* (London, 1678), p. 65.

44. This result was first published (with a photograph) in Lawrence M. Principe, "Apparatus and Reproducibility in Alchemy," in *Instruments and Experimentation in the History of Chemistry*, ed. Frederic L. Holmes and Trevor Levere, (Cambridge, MA: MIT Press, 2000), pp. 55–74, the proceedings of a conference held at the Dibner Institute at MIT in April 1996.

45. Starkey, *Alchemical Laboratory Notebooks and Correspondence*, pp. 84–85.

46. Ibid., p. 21; Boyle, *Correspondence*, 1:95.

47. Lemery, *Cours de chymie* (Paris, 1675), pp. 68–69.

48. Ewald van Hoghelande, *Historiae aliquot transmutationis metallicae . . . pro defensione alchymiae contra hostium rabiem* (Cologne, 1604); Siegmund Heinrich Güldenfalk, *Sammlung von mehr als hundert wahrhaftigen Transmutationsgeschichten* (Frankfurt, 1784), in relation to which, see also Jürgen Strein, "Siegmund Heinrich Güldenfalks *Sammlung von mehr als 100 Transmutationsgeschichten* (1784)," in *Iliaster: Literatur und Naturkunde in der frühen Neuzeit*, ed. Wilhelm Kühlmann and Wolf-Dieter

Müller-Jahncke (Heidelberg: Manutius Verlag, 1999), pp. 275–83; Bernard Husson, *Transmutations alchimiques* (Paris: Editions J'ai Lu, 1974).

49. The story is told in detail by Klaus Hoffmann, *Johann Friedrich Böttger: Vom Alchemistengold zum weissen Porzellan* (Berlin: Verlag Neues Leben, 1985), and more loosely in Janet Gleeson, *The Arcanum: The Extraordinary True Story* (New York: Warner, 1998); for a contemporary account, see Gottfried Wilhelm Leibniz, "Oedipus chymicus," *Miscellanea Berolinensia* 1 (1710): 16–21.

50. One (notorious) example is Wenzel Seyler; see Pamela Smith, "Alchemy as a Language of Mediation in the Habsburg Court," *Isis* 85 (1994):1–25. For more on Seyler, see Johann Joachim Becher, *Magnalia naturae* (London, 1680), which contains a moralizing account of Seyler's discovery, theft, and abuse of the Philosophers' Stone and his subsequent adventures at court. Other contemporary accounts of him are published in Principe, *Aspiring Adept*, pp. 261–63 and 296–300.

51. Samuel Reyher, *Dissertatio de nummis quibusdam ex chymico metallo factis* (Kiel, Germany, 1690); for modern studies of these coins, see Vladimir Karpenko, "Coins and Medals Made of Alchemical Metal," *Ambix* 35 (1988): 65–76; "Alchemistische Münzen und Medaillen," in *Anzeiger der Germanischen Nationalmuseums 2001* (Nuremberg: Germanisches Nationalmuseum, 2001), pp. 49–72; and *Alchemical Coins and Medals* (Glasgow: Adam Maclean, 1998).

52. Johann Friedrich Helvetius, *Vitulus aureus* (Amsterdam, 1667). It was published at Nuremberg in German (*Guldenes Kalb*) in the next year, at London in English (*The Golden Calf*) in 1670, and reprinted in *Musaeum hermeticum* in 1678, pp. 815–63.

53. Ibid., p. 894.

54. Benedict Spinoza, *Spinoza Opera im Auftrag der Heidelberger Akademie der Wissenschaften*, ed. Carl Gebhardt (Heidelberg, [1925]), vol. 4, *Epistolae*, pp. 196–97.

55. Boyle's *Dialogue* remained unpublished and scattered in pieces among his voluminous papers until the 1990s. It is now published in Principe, *Aspiring Adept*, pp. 223–95.

56. Robert Boyle, *Dialogue on Transmutation*, in Principe, *Aspiring Adept*, p. 265; the spelling and punctuation is partly modernized here for clarity.

57. Ibid.

58. Ibid., p. 266.

59. Ibid., p. 268.

60. "Burnet Memorandum," printed in Michael Hunter, *Robert Boyle by Himself and His Friends* (London: Pickering, 1994), p. 30. In fact, this event was only the most dramatic and certain of Boyle's several witnesses of metallic transmutation.

61. See Hunter, "Alchemy, Magic, and Moralism," esp. p. 405.

Chapter Seven

1. On Maier, see Erik Leibenguth, *Hermetische Poesie des Frühbarock: Die "Cantilenae intellectuales" Michael Maiers* (Tübingen: Max Niemeyer Verlag, 2002); Karin Figala and Ulrich Neumann, "'Author, Cui Nomen Hermes Malavici': New Light on the Biobibliography of Michael Maier (1569–1622)," in Rattansi and Clericuzio, *Alchemy and Chemistry in the Sixteenth and Seventeenth Centuries*, pp. 121–48, and "À propos de Michel Maier: Quelques découvertes bio-bibliographiques," in Kahn and Matton, *Alchimie*, pp. 651–61; and Ulrich Neumann, "Michel Maier (1569–1622): Philosophe et médecin," in Margolin and Matton, *Alchimie et philosophie à la*

Renaissance, pp. 307–26. For an older study, now dated, in English see J. B. Craven, *Count Michael Maier, Doctor of Philosophy and Medicine, Alchemist, Rosicrucian, Mystic, 1568–1622* (Kirkwall, UK: Peace and Sons, 1910).

2. The identification of Maier's sources has been carried out by H. M. E. de Jong, *Michael Maier's Atalanta Fugiens: Sources of an Alchemical Book of Emblems* (Leiden: Brill, 1969). There are many other emblematic books of the florilegium type as well; for example, the *Viridarium chymicum* of Daniel Stoltzius von Stoltzenberg (Frankfurt, 1624), also published in a German version, the *Chymisches Lustgärtlein* (Frankfurt, 1624; reprint, Darmstadt: Wissenschaftliche Buchgesellschaft, 1964).

3. One version of the story is told in Ovid, *Metamorphoses*, 10:560–707.

4. For example, emblem 24 is clearly dependent on Basil Valentine's first key, but Maier adds his own hint for the reader. The last line of the epigram reads: "And the king shall be proud with a leonine heart." The phrase *leonino corde* should remind readers with a knowledge of astronomy of the celestial *Cor leonis*, the brightest star in Leo, and known today as α Leonis. But the star has another name: Regulus, literally meaning "little king." But in chymistry, *regulus* means specifically a metallic alloy containing antimony—in other words, *exactly* the substance produced during the chemical process encoded in the emblem.

5. See Jacques Rebotier, "La musique cachée de l'*Atalanta fugiens*," *Chrysopoeia* 1 (1987): 56–76; for more on music in alchemy, see Christoph Meinel, "Alchemie und Musik," in Meinel, *Die Alchemie in der europäischer Kultur- und Wissenschaftsgeschichte*, pp. 201–28; and Jacques Rebotier, "La *Musique de Flamel*," in Kahn and Matton, *Alchimie*, pp. 507–46.

6. On the place and content of emblems, see the outstanding study by John Manning, *The Emblem* (London: Reaktion Books, 2002), and references therein. It is worth noting that Alciati originally composed only allusive poems; the allegorical emblems were probably added later. See ibid., pp. 38–43. Unfortunately, contemporary studies of humanist and literary emblems rarely make more than passing reference to chymical emblems, and vice versa. See also Alison Adams and Stanton J. Linden, eds., *Emblems and Alchemy* (Glasgow: Glasgow Emblem Studies, 1998), although the quality of the contributed chapters therein is uneven.

7. Michael Maier, *Atalanta fugiens* (Oppenheim, 1618), p. 6.

8. On such issues see Jean-Marc Mandosio, "La place de l'alchimie dans les classifications des sciences et des arts à la Renaissance," *Chrysopoeia* 4 (1990–91): 199–282, and Sylvain Matton, "L'influence de l'humanisme sur la tradition alchimique," in "Le crisi dell'alchemia," *Micrologus* 3 (1995): 279–345.

9. Dante, *La divina commedia*, canto 29.

10. Petrarch, *Remedies for Fortune Fair and Foul*, trans. Conrad H. Rawski (Bloomington: Indiana University Press, 1991), 1:299–301.

11. Georgius Agricola, *De re metallica* (Basel, 1556). For Agricola's biography, see H. M. Wilsdorf, *Georg Agricola und seine Zeit* (Berlin: Deutsche Verlag der Wissenschaften, 1956), and Hans Prescher, *Georgius Agricola: Persönlichkeit und Wirken für den Bergbau und das Hüttenwesen des 16. Jahrhunderts* (Weinheim: VCH, 1985). For his humanist training and program, see Owen Hannaway, "Georgius Agricola as Humanist," *Journal of the History of Ideas* 53 (1992): 553–60.

12. See Zweder van Martels, "Augurello's *Chrysopoeia* (1515): A Turning Point in the Literary Tradition of Alchemical Texts," *Early Science and Medicine* 5 (2000): 178–95.

13. Michael Maier, *Arcana arcanissima* (London, 1613). Many other chymical books from the sixteenth to the eighteenth century develop the theme; see for example Vincenzo Percolla, *Auriloquio*, ed. Carlo Alberto Anzuini, Textes et Travaux de Chrysopoeia 2 (Paris: SÉHA; Milan: Archè, 1996); Pierre-Jean Fabre, *Hercules piochymicus* (Toulouse, 1634); and Antoine-Joseph Pernety, *Les fables égyptiennes et grecques dévoilées* (Paris, 1758; reprint, with a useful introduction by Sylvain Matton, Paris: La Table d'émeraude, 1982), and *Dictionnaire mytho-hermétique* (Paris, 1758). One of the earliest alchemical interpretations of Greek myths occurs briefly in Petrus Bonus's early fourteenth-century *Margarita preciosa novella*, in *Bibliotheca chemica curiosa*, 2:1–80, on 42–43. For a modern analysis of the topic, see Sylvain Matton, "L'interprétation alchimique de la mythologie," *Dix-huitième siècle* 27 (1995):73–87.

14. Sylvain Matton, "Une lecture alchimique de la Bible: Les 'Paradoxes chimiques' de Francois Thybourel," *Chrysopoeia* 2 (1988): 401–22; Didier Kahn, "L'interprétation alchimique de la Genèse chez Joseph Du Chesne dans le contexte de ses doctrines alchimiques et cosmologiques," pp. 641–692 in *Scientiae et artes: Die Vermittlung alten und neuen Wissens in Literatur, Kunst und Musik*, ed. Barbara Mahlmann-Bauer (Wiesbaden: Harrassowitz, 2004), pp. 641–92; Peter J. Forshaw, "Vitriolic Reactions: Orthodox Responses to the Alchemical Exegesis of Genesis," in *The Word and the World: Biblical Exegesis and Early Modern Science*, ed. Kevin Killeen and Peter J. Forshaw (Basingstoke: Palgrave, 2007), pp. 111–36.

15. Thomas Sprat, *A History of the Royal Society of London* (London, 1667), p. 37.

16. Herman Boerhaave, *Sermo academicus de chemia suos errores expurgante* (Leiden, 1718), reprinted in his *Elementa chemiae* 2:64–77, quoting from p. 66; for an English translation, see E. Kegel-Brinkgreve and Antonie M. Luyendijk-Elshout, eds., *Boerhaave's Orations* (Leiden: Brill, 1983), pp. 193–213, quoting from p. 195. On Boerhaave, see John C. Powers, *Inventing Chemistry: Herman Boerhaave and the Reform of the Chemical Arts* (Chicago: University of Chicago Press, 2012).

17. Exodus 32:20; 1 Kings 9:28; 2 Chronicles 8:18.

18. Zosimos, unfortunately, was not ancient enough, having lived in the last days of a degraded Roman Empire. Most humanists displayed indifference to the *Corpus alchemicum graecum* with its inelegant Greek, although supporters of chrysopoeia stressed it as part of their early pedigree. See Matton, "L'influence," pp. 309–41.

19. See Robert Halleux, "La Controverse sur les origines de la chimie de Paracelse à Borrichius," in *Acta conventus neo-latini Turonensis* (Paris: Vrin, 1980), 2:807–17, esp. p. 809.

20. For arguments over the antiquity of chymistry, see for example Olaus Borrichius, *De ortu et progressu chemiae* (Copenhagen, 1668), reprinted in *Biblioteca chemica curiosa*, 1:1–37, and *Hermetis, Aegyptiorum et chemicorum sapientia ab Hermanni Conringii animadversionibus vindicata* (Copenhagen, 1674); and Hermann Conring, *De Hermetica Aegyptorum* (Helmstadt, 1648) and *De Hermetica medicina* (Helmstadt, 1669 [enlarged edition of the 1648 publication]). On the debate concerning the *prisca sapientia*, see Martin Muslow, "Ambiguities of the *Prisca Sapientia* in Late Renaissance Humanism," *Journal of the History of Ideas* 65 (2004): 1–13. On Newton's interest, see McGuire and Rattansi, "Newton and the Pipes of Pan."

21. Quoted from the modernized version: Geoffrey Chaucer, *Canterbury Tales*, trans. David Wright (Oxford: Oxford University Press, 1985), pp. 449–50; for the original, see Chaucer, *The Canon's Yeoman's Tale*, ed. Maurice Hussey (Cambridge: Cambridge University Press, 1965), pp. 53–54 (lines 889–94 and 919–23).

22. Edgar H. Duncan, "The Literature of Alchemy and Chaucer's Canon's Yeoman's Tale: Framework, Theme, and Characters," *Speculum* 43 (1968): 633–56; "The Yeoman's Canon's 'Silver Citrinacioun,'" *Modern Philology* 37 (1940): 241–62. Elias Ashmole printed the "Canon's Yeoman's Tale" in his collection of alchemical poetry, *Theatrum chemicum britannicum*, pp. 227–56, and writes that one reason he included it was "to shew that *Chaucer* himselfe was a *Master*" in alchemy (p. 467).

23. Thomas Norton, *Ordinall of Alchimy*, in Ashmole, *Theatrum chemicum britannicum*, p. 7.

24. For example, the painting shown in plate 2 does accurately depict the kind of explosions that were a regular occurrence in contemporaneous chymical laboratories; however, the real message of the painting lies in the background, in the scatological commentary wordlessly expressed by a woman wiping her child's bottom.

25. For interpretations of this print, and of alchemy in art more generally, see Lawrence M. Principe and Lloyd Dewitt, *Transmutations: Alchemy in Art* (Philadelphia: Chemical Heritage Foundation, 2002), pp. 11–12, and A. A. A. M. Brinkman, *De Alchemist in de Prentkunst* (Amsterdam: Rodopi, 1982), pp. 41–53. See also Jane Russell Corbett, "Conventions and Change in Seventeenth-Century Depictions of Alchemists," in *Alchemy and Art*, ed. Jacob Wamberg (Copenhagen: Museum Tusculanum Press, 2006), pp. 249–71, and A. A. A. M. Brinkman, *Chemie in de Kunst* (Amsterdam: Rodopi, 1975). The "classic" work on the subject is Jacques van Lennep, *Art et alchimie* (Brussels: Meddens, 1966). However, although the book contains useful inventories, its interpretations are based on now-outdated notions about alchemy, and so should be approached with great caution.

26. For the linkage of alchemy to Elizabethan and Stuart theatre and literary genres, see Stanton J. Linden, *Darke Hieroglyphicks: Alchemy in English Literature from Chaucer to the Restoration* (Lexington: University Press of Kentucky, 1996).

27. Edgar Hall Duncan, "Jonson's *Alchemist* and the Literature of Alchemy," *Proceedings of the Modern Language Association* 61 (1946): 699–710; "The Alchemy in Jonson's *Mercury Vindicated*," *Studies in Philology* 39 (1942): 625–37; Stanton J. Linden, "Jonson and Sendivogius: Some New Light on 'Mercury Vindicated,'" *Ambix* 24 (1977): 39–54.

28. The "day of Projection" refers to the momentous day on which the chymist tests the transmutatory ability of his Philosophers' Stone for the first time. William Congreve, *Way of the World*, in *The Complete Plays of William Congreve*, ed. Herbert Davis (Chicago: University of Chicago Press, 1967), pp. 46 and 431.

29. "Que la chimie est admirable," from [Michel Chilliat?], *Les Souffleurs; ou, La pierre philosophale d'Arlequin* (Paris, 1694), pp. 114–15 and 121. The first edition contains nine pieces of music written for the play, although most copies lack all or most of the scores. For more on alchemy in French and Italian theatre, see the contributed essays in "Thèâtre et Alchimie," *Chrysopoeia* 2 (1988), fascicle 1.

30. John Donne, "Loves Alchymie," in *The Complete English Poems of John Donne*, ed. C. A. Patrides (London: J. M. Dent, 1985), p. 86. On Donne and chymistry, see Jocelyn Emerson, "John Donne and the Noble Art," in *Textual Healing: Essays in Medieval and Early Modern Medicine*, ed. Elizabeth Lane Furdell (Leiden: Brill, 2005), pp. 195–221, and Edgar Hill Duncan, "Donne's Alchemical Figures," *English Literary History* 9 (1942): 257–85.

31. Two examples are Michael Maier, *Examen fucorum pseudo-chymicorum detectorum et in gratiam veritatis amantium succincte refutatorum* (Frankfurt, 1617), and

Heinrich Khunrath, *Trewhertzige Warnungs-Vermahnung* (Magdeburg, 1597). On the former, see Wolfgang Beck, "Michael Maiers Examen Fucorum Pseudo-chymicorum: eine Schrift wider die falschen Alchemisten, " PhD diss., Technische Universität München, 1992, and Robert Halleux, "L'alchimiste et l'essayeur," in Meinel, *Die Alchemie in der europaischen Kultur- und Wissenschaftsgeschichte*, pp. 277–91.

32. This term and an archivally rich study of "contractual alchemy" and the construction of the category the "false alchemist" are presented in Tara Nummedal, *Alchemy and Authority in the Holy Roman Empire* (Chicago: University of Chicago Press, 2007).

33. For example, see William Eamon, "Alchemy in Popular Culture: Leonardo Fioravanti and the Search for the Philosopher's Stone," *Early Science and Medicine* 5 (2000): 196–213, and Tara Nummedal, "Words and Works in the History of Alchemy," *Isis* 102 (2011): 330–37.

34. Bruce Moran, *The Alchemical World of the German Court*, Sudhoffs Archiv 29 (Stuttgart: Steiner Verlag, 1991); Pamela H. Smith, *The Business of Alchemy: Science and Culture in the Holy Roman Empire* (Princeton, NJ: Princeton University Press, 1994); Jost Weyer, *Graf Wolfgang von Hohenlohe und die Alchemie: Alchemistische Studien in Schloss Weikersheim 1587–1610* (Sigmaringen, Germany: Thorbecke, 1992); Mar Rey Bueno, *Los señores del fuego: Destiladores y espagíricos en la corte de los Austrias* (Madrid: Corona Borealis, 2002), and "La alquimia en la corte de Carlos II (1661–1700)," *Azogue* 3 (2000), online at http://www.revistaazogue.com; Alfredo Perifano, "Theorica et practica dans un manuscrit alchimique de Sisto de Boni Sexti da Norcia, alchimiste à la cour de Côme Ier de Médicis," *Chrysopoeia* 4 (1990–91): 81–146; Didier Kahn, "King Henry IV, Alchemy, and Paracelsianism in France (1589–1610)," in Principe, *Chymists and Chymistry*, pp. 1–11.

35. On this topic see Sylvain Matton, "Thématique alchimique et litterature religieuse dans la France du XVIIᵉ siècle," *Chrysopoeia* 2 (1998): 129–208; Matton, *Scolastique et alchimie*, pp. 661–737; and Sylvia Fabrizio-Costa, "De quelques emplois des thèmes alchimiques dans l'art oratoire italien du XVIIᵉ siècle," *Chrysopoeia* 3 (1989): 135–62.

36. For example, 1 Peter 1:7, Proverbs 17:3 and 27:21, Wisdom 3:6, and Job 23:10.

37. See Sylvain Matton, "Remarques sur l'alchimie transmutatoire chez les théologiens réformés de la Renaissance," *Chrysopoeia* 7 (2000–2003): 171–87, esp. pp. 172–75.

38. Jean-Pierre Camus, cited in Matton, "Thématique alchimique," p. 149.

39. Matton, "Remarques sur l'alchimie transmutatoire"; John Slater, "Rereading Cabriada's *Carta*: Alchemy and Rhetoric in Baroque Spain," *Colorado Review of Hispanic Studies* 7 (2009): 67–80, esp. 73–75.

40. Cited in Matton, *Scolastique et alchimie*, p. 726.

41. Norton, *Ordinall*, p. 13; on Norton, see J. Reidy, "Thomas Norton and the *Ordinall of Alchimy*," *Ambix* 6 (1957): 59–85.

42. The manuscript is British Library, Additional MS 10302. Its illumination differs in intriguing ways from the later printed engraving.

43. St. Thomas Aquinas, *Summa theologica*, 1ae 2a, quaestio 112, articulus 5 and 2ae 2a, quaestio 9; for further references and a treatment of this issue in legal and ethical contexts, see Gaines Post, Kimon Giocarinis, and Richard Kay, "The Medieval Heritage of a Humanistic Ideal: 'Scientia donum dei est, unde vendi non potest,'" *Traditio* 11 (1955): 195–234, and Gaines Post, "Master's Salaries and Student-Fees in Mediaeval Universities," *Speculum* 7 (1932): 181–98.

44. Note that proscriptions against putting a price tag on knowledge did not imply that it was to be given away freely to everyone, regardless of the legal precept's connection to the divine command "freely you have received, freely give" (Matthew 10:8). On this issue, see Carla Hesse, "The Rise of Intellectual Property, 700 BC–AD 2000: An Idea in the Balance," *Daedalus* 131 (2002): 26–45.

45. Starkey, *Alchemical Laboratory Notebooks and Correspondence*, p. 175.

46. For Starkey's frequent acknowledgements of divine assistance, see ibid., pp. 43, 67–69, 113, 190, and 302; for more treatment of this issue, see Newman and Principe, *Alchemy Tried in the Fire*, pp. 197–205.

47. For a brief presentation of the foundations and implications of the "connected world" of the early modern, see Principe, *The Scientific Revolution*, pp. 21–38.

48. For the literature on Fludd, see the useful bibliographical essay in Allen G. Debus, ed., *Robert Fludd and His Philosophical Key* (New York: Science History Publications, 1979), pp. 51–52; for a more recent study, Johannes Rösche, *Robert Fludd: Der Versuch einer hermetischen Alternative zur neuzeitlichen Naturwissenschaft* (Göttingen: V&R, 2008); for more on alchemy specifically, see also François Fabre, "Robert Fludd et l'alchimie: Le *Tractatus Apologeticus integritatem societatis de Rosea Cruce defendens*," *Chrysopoeia* 7 (2000–2003): 251–91. For Fludd's link to Harvey, see Allen G. Debus, "Robert Fludd and the Circulation of the Blood," *Journal of the History of Medicine and Related Sciences* 16 (1961): 374–93.

49. Much unreliable material has been written about Khunrath; for solid scholarship, see the work currently being produced by Peter Forshaw, such as "Alchemy in the Amphitheatre: Some Considerations of the Alchemical Content of the Engravings in Heinrich Khunrath's *Amphitheatre of Eternal Wisdom* (1609)," in Wamberg, *Alchemy and Art*, pp. 195–220.

50. Heinrich Khunrath, *Lux in tenebris* (n.p., 1614), pp. 3–4.

51. Ibid., pp. 9–10.

52. See chapter 3, p. 68.

53. Robert Boyle, *Some Physico-Theological Considerations about the Possibility of the Resurrection*, in *Works of Robert Boyle*, 8:295–327. A similar usage actually dates to the fifth-century Aeneas of Gaza, *Theophrastus*, in *Patrologia graeca*, ed. J. P. Migne (Paris, 1868), 85:871–1003, esp. 983–84 and 992. See also Matton, "Thématique alchimique," pp. 180–90.

54. Thomas Browne, *Religio medici*, in *Works of Sir Thomas Browne*, ed. Geoffroy Keynes (Chicago: University of Chicago Press, 1964) 1:50.

55. Pierre-Jean Fabre, *Alchymista christianus* (Tolouse, 1632); available as a reprint with accompanying French translation as *L'alchimiste chrétien*, ed. and trans. Frank Grenier, Textes et Travaux de Chrysopoeia 7 (Paris: SÉHA; Milan: Archè, 2001). The quotations are excerpts from the full title. For Fabre's *Manuscriptum ad Fredericum*, an orderly, explanatory text extremely useful for gaining a fuller appreciation of chrysopoetic theory and principles in the seventeenth century, see the edition, French translation, and exposition in Bernard Joly, *La rationalité de l'alchimie au XVIIᵉ siècle* (Paris: Vrin, 1992).

56. Johannes Kepler, *Epitome of Copernican Astronomy*, bk. 4 in *Ptolemy, Copernicus, Kepler; Britannica Great Books*, vol. 16 (Chicago: Encyclopedia Britannica, 1952), pp. 853–54.

57. Johannes Kepler, *Harmonices mundi*, in *Gesammelte Werke*, ed. Max Caspar (Munich: Beck Verlag, 1940) 6:366; in English in *Ptolemy, Copernicus, Kepler*, 16:1083.

58. Mark A. Waddell, "Theatres of the Unseen: The Society of Jesus and the Problem of the Invisible in the Seventeenth Century" (PhD diss., Johns Hopkins University, 2006), pp. 80–114.

59. For two definitions of natural philosophy, see Walter Pagel, "The Vindication of Rubbish," originally published in the 1945 *Middlesex Hospital Journal*, reprinted in *Religion and Neoplatonism in Renaissance Medicine* (London: Variorum, 1985), 1–14, on p. 11, and Dennis Des Chene, *Physiologia: Natural Philosophy in Late Aristotelian and Cartesian Thought* (Ithaca, NY: Cornell University Press, 1996), p. 3: a topic "in which physics, metaphysics, and theology could meet and negotiate their claims."

❄ BIBLIOGRAPHY ❄

Abrahams, Harold J. "Al-Jawbari on False Alchemists." *Ambix* 31 (1984): 84–87.

Adelung, Johann Christoph. *Geschichte der menschlichen Narrheit; oder, Lebensbeschreibungen berühmter Schwarzkünstler, Goldmacher, Teufelsbanner, Zeichen- und Liniendeuter, Schwärmer, Wahrsager, und anderer philosophischer Unholden.* 7 vols. Leipzig, 1785–89.

Agricola, Georgius. *De re metallica.* Basel, 1556.

Albert the Great. *Alberti Magni opera omnia.* Edited by August Borgnet. 37 vols. Paris, 1890–99.

———. *"Libellus de alchimia" Ascribed to Albertus Magnus.* Translated by Virginia Heines, SCN. Berkeley: University of California Press, 1958.

Al-Jawbari. *La voile arraché.* Translated by René R. Khawan. 2 vols. Paris: Phèbus, 1979.

Anawati, Georges C. "L'alchimie arabe." In Rashed and Morelon, *Histoire des sciences arabes,* 3:111–42.

———. "Avicenna et l'alchimie." In *Convegno internazionale, 9–15 aprile 1969: Oriente e occidente nel medioevo; filosofia e scienze,* pp. 285–345. Rome: Accademia Nazionale dei Lincei, 1971.

Anthony, Francis. *The apologie, or defence of . . . aurum potabile.* London, 1616.

Arnald of Villanova, pseudo-. *De secretis naturae.* Edited and translated by Antoine Calvet. In "Cinq traités alchimique médiévaux," *Chrysopoeia* 6 (1997–99): 154–206.

———. *Thesaurus thesaurorum et rosarium philosophorum.* In *Bibliotheca chemica curiosa,* 1:662–76.

———. *Tractatus parabolicus.* Edited and translated by Antoine Calvet. *Chrysopoeia* 5 (1992–96): 145–71.

Ashmole, Elias, ed. *Theatrum chemicum britannicum.* London, 1652.

Atwood, Mary Anne. *A Suggestive Inquiry into the Hermetic Mystery.* London: T. Saunders, 1850. Reprint, Belfast: William Tait, 1918.

Aurnhammer, Achim. "Zum Hermaphroditen in der Sinnbildkunst der Alchemisten." In Meinel, *Die Alchemie in der europäischen Kultur- und Wissenschaftsgeschichte*, pp. 179–200.

Avicenna. *See* Ibn-Sīnā.

Bagliani, Agostino Paravicini. "Ruggero Bacone e l'alchimia di lunga vita: Riflessioni sui testi." In "Alchimia e medicina nel Medioevo," *Micrologus* 9 (2003): 33–54.

Baldwin, Martha. "Alchemy and the Society of Jesus in the Seventeenth Century: Strange Bedfellows?" *Ambix* 40 (1993): 41–64.

Balīnūs. "Le *De secretis naturae* du pseudo-Apollonius de Tyane: Traduction latine par Hugues de Santalla du *Kitāb sirr al-ḫalīqa* de Balīnūs." Edited by Françoise Hudry. In "Cinq traités alchimique médiévaux," *Chrysopoeia* 6 (1997–99): 1–153.

———. *Sirr al-khalīqah wa ṣanʿāt al-ṭabīʿah*. Edited by Ursula Weisser. Aleppo: Aleppo Institute for the History of Arabic Science, 1979.

Baud, Jean-Pierre. *Le procès d'alchimie*. Strasbourg: CERDIC, 1983.

Baudrimont, Alexandre. *Traité de chimie générale et expérimentale*. Paris, 1844.

Becher, Johann Joachim. *Magnalia naturae*. London, 1680.

Beck, Wolfgang. "Michael Maiers Examen Fucorum Pseudo-chymicorum: Eine Schrift wider die falschen Alchemisten. " PhD diss., Technische Universität München, 1992.

Beguin, Jean. *Tyrocinium chymicum*. Paris, 1612.

Benson, Robert L., and Giles Constable, eds. *Renaissance and Renewal in the Twelfth Century*. With Carol D. Lanham. Cambridge, MA: Harvard University Press, 1982. Reprint, Toronto: Medieval Academy of America, 1991.

Benzenhöfer, Udo. *Johannes' de Rupescissa Liber de consideratione quintae essentiae omnium rerum deutsch*. Stuttgart: Franz Steiner Verlag, 1989.

———. *Paracelsus*. Reinbek: Rowohlt, 1997.

Beretta, Marco. *The Alchemy of Glass: Counterfeit, Imitation, and Transmutation in Ancient Glassmaking*. Sagamore Beach, MA: Science History Publications, 2009.

Berthelot, Marcellin. *La chimie au moyen âge*. 3 vols. Paris, 1893.

Berthelot, Marcellin, and C. E. Ruelle, eds. *Collections des alchimistes grecs*. 3 vols. Paris, 1888.

Bibliotheca chemica curiosa. Edited by J. J. Manget. 2 vols. Geneva, 1702. Reprint, Sala Bolognese: Arnoldo Forni, 1976.

Bidez, Joseph et al., eds. *Catalogue des manuscrits alchimiques grecs*. 8 vols. Brussels: Lamertin, 1924–32.

Bignami-Odier, Jeanne. "Jean de Roquetaillade." In *Histoire littéraire de la France*, 41:75–240.

Boerhaave, Herman. *Elementa chemiae*. 2 vols. Paris, 1733.

Bolton, H. Carrington. "Hysterical Chemistry." *Chemical News* 77 (1898): 3–5, 16–18.

———. "The Revival of Alchemy." *Science* 6 (1897): 853–63.

Bonus, Petrus. *Margarita preciosa novella*. In *Bibliotheca chemica curiosa*, 2:1–80.

Borrichius, Olaus. *Conspectus scriptorum chemicorum celebriorum*. In *Bibliotheca chemica curiosa*, 1:38–53.

———. *De ortu et progressu chemiae*. Copenhagen, 1668. Reprinted in *Bibliotheca chemica curiosa*, 1:1–37.

———. *Hermetis, Aegyptiorum et chemicorum sapientia ab Hermanni Conringii animadversionibus vindicata*. Copenhagen, 1674.

Bouyer, Louis. "Mysticism: An Essay on the History of a Word." In *Understanding Mysticism*, pp. 42–55. Garden City, NY: Image Books, 1980.

Boyle, Robert. *The Correspondence of Robert Boyle.* Edited by Michael Hunter, Lawrence M. Principe, and Antonio Clericuzio. 6 vols. London: Pickering and Chatto, 2001.

———. *Dialogue on Transmutation.* Edited in Principe, *The Aspiring Adept*, pp. 233–95.

———. *The Works of Robert Boyle.* Edited by Michael Hunter and Edward B. Davis. 14 vols. London: Pickering and Chatto, 1999–2000.

Brinkman, A. A. A. M. *De Alchemist in de Prentkunst.* Amsterdam: Rodopi, 1982.

———. *Chemie in de Kunst.* Amsterdam: Rodopi, 1975.

Brunschwig, Jacques, and Geoffrey E. R. Lloyd, eds. *Greek Thought: A Guide to Classical Knowledge.* Cambridge, MA: Belknap Press of Harvard University Press, 2000.

Buddeus, Johann Franz. *Quaestionem Politicam an alchimistae sint in republica tolerandi?* Magdeburg, 1702. Translated in German under the title *Untersuchung von der Alchemie*; in *Deutsches Theatrum Chemicum*, 1:1–146.

Buntz, Herwig. "Das *Buch der heiligen Dreifaltigkeit*, sein Autor und seine Überlieferung." *Zeitschrift für deutsches Altertums und deutsche Literatur* 101 (1972): 150–60.

Burkhalter, Fabienne. "La production des objets en métal (or, argent, bronze) en Égypte Hellénistique et Romaine à travers les sources papyrologiques." In *Commerce et artisanat dans l'Alexandrie hellénistique et romaine*, edited by Jean-Yves Empereur, pp. 125–33. Athens: EFA, 1998.

Burr, David. *The Spiritual Franciscans: From Protest to Persecution in the Century after St. Francis.* University Park: Penn State University Press, 2001.

Caley, Earle Radcliffe. "The Leiden Papyrus X: An English Translation with Brief Notes." *Journal of Chemical Education* 3 (1926): 1149–66.

———. "The Stockholm Papyrus: An English Translation with Brief Notes." *Journal of Chemical Education* 4 (1927): 979–1002.

Calvet, Antoine. "Alchimie et Joachimisme dans les *alchimica* pseudo-Arnaldiens." In Margolin and Matton, *Alchimie et philosophie à la Renaissance*, pp. 93–107.

———. "Un commentaire alchimique du XIVᵉ siècle: Le *Tractatus parabolicus* du ps.-Arnaud de Villaneuve." In *Le Commentaire: Entre tradition et innovation*, edited by Marie-Odile Goulet-Cazé, pp. 465–74. Paris: Vrin, 2000.

———. "Étude d'un texte alchimique latin du XIVᵉ siècle: Le *Rosarius philosophorum* attribué au medecin Arnaud de Villeneuve." *Early Science and Medicine* 11 (2006): 162–206.

———. "La théorie *per minima* dans les textes alchimiques des XIVe et XVᵉ siècles." In López-Pérez, Kahn, and Rey Bueno, *Chymia*, pp. 41–69.

Cambriel, L. P. François. *Cours de philosophie hermétique ou d'alchimie.* Paris, 1843.

Cameron, H. Charles. "The Last of the Alchemists." *Notes and Records of the Royal Society* 9 (1951): 109–14.

Caron, Richard. "Notes sur l'histoire de l'alchimie en France à la fin du XIXᵉ et au début du XXᵉ siècle." In *Ésotérisme, gnoses & imaginaire symbolique*, edited by Richard Caron, Joscelyn Godwin, Wouter J. Hanegraaff, and Jean-Louis Vieillard-Baron, pp. 17–26. Leuven: Peeters, 2001.

Casaubon, Meric. *A True and Faithfull Relation.* London, 1659.

Chang, Ku-Ming (Kevin). "The Great Philosophical Work: Georg Ernst Stahl's Early Alchemical Teaching." In López-Pérez, Kahn, and Rey Bueno, *Chymia*, pp. 386–96.

————. "Toleration of Alchemists as Political Question: Transmutation, Disputation, and Early Modern Scholarship on Alchemy." *Ambix* 54 (2007): 245–73.

Chaucer, Geoffrey. *Canterbury Tales*. Translated by David Wright. Oxford: Oxford University Press, 1985.

[Chilliat, Michel?]. *Les Souffleurs; ou, La pierre philosophale d'Arlequin*. Paris, 1694.

Cockren, Archibald. *Alchemy Rediscovered and Restored*. London: Rider, 1940.

Coelum philosophorum. Frankfurt and Leipzig, 1739.

Cohen, I. Bernard. "Ethan Allen Hitchcock: Soldier-Humanitarian-Scholar, Discoverer of the 'True Subject' of the Hermetic Art." *Proceedings of the American Antiquarian Society* 61 (1951): 29–136.

Collectanea chymica. London, 1893.

Collesson, Jean. *Idea perfecta philosophiae hermeticae*. In *Theatrum chemicum*, 6:143–62.

Congreve, William. *The Complete Plays*. Edited by Herbert Davis. Chicago: University of Chicago Press, 1967.

Conring, Hermann. *De Hermetica Aegyptorum*. Helmstadt, 1648.

————. *De Hermetica medicina*. Helmstadt, 1669.

Constantine of Pisa. *The Book of the Secrets of Alchemy*. Edited and translated by Barbara Obrist. Leiden: Brill, 1990.

Copenhaver, Brian. *Hermetica: The Greek Corpus Hermeticum and the Latin Asclepius*. Cambridge: Cambridge University Press, 1992.

Corbett, Jane Russell. "Conventions and Change in Seventeenth-Century Depictions of Alchemists." In Wamberg, *Alchemy and Art*, pp. 249–71.

Craven, J. B. *Count Michael Maier, Doctor of Philosophy and Medicine, Alchemist, Rosicrucian, Mystic, 1568–1622*. Kirkwall, UK: Peace and Sons, 1910.

Cremer, Abbot. *Testamentum Cremeri*. In *Musaeum hermeticum*, pp. 531–44.

Crisciani, Chiara. "Exemplum Christi e sapere: Sull'epistemologia di Arnoldo da Villanova." *Archives internationales d'histoire des sciences* 28 (1978): 245–87.

————. *Il Papa e l'alchimia: Felice V, Guglielmo Fabri e l'elixir*. Rome: Viella, 2002.

Crisciani, Chiara, and Agostino Paravicini Bagliani, eds. *Alchimia e medicina nel Medioevo*. Micrologus Library 9. Florence: Sismel, 2003.

Cunningham, Andrew. "Paracelsus Fat and Thin: Thoughts on Reputations and Realities," In Grell, *Paracelsus*, pp. 53–77.

Cyliani. *Hermés dévoilé*. Paris, 1832. Reprint, Paris: Éditions Traditionnelles, 1975.

Darmstaedter, Ernst. "Zur Geschichte des *Aurum potabile*." *Chemiker-Zeitung* 48 (1924): 653–55, 678–80.

————. "Liber Misericordiae Geber: Eine lateinische Übersetzung des grösseren Kitāb alrahma." *Archiv für Geschichte der Medizin* 17 (1925): 187–97.

De auro potabili. In *Theatrum chemicum*, 6:382–93.

Debus, Allen G. *The Chemical Philosophy: Paracelsian Science and Medicine in the Sixteenth and Seventeenth Centuries*. 2 vols. New York: Science History Publications, 1977.

————. *The French Paracelsians*. Cambridge: Cambridge University Press, 1991.

————. "Robert Fludd and the Circulation of the Blood." *Journal of the History of Medicine and Related Sciences* 16 (1961): 374–93.

————. *Robert Fludd and His Philosophical Key*. New York: Science History Publications, 1979.

Del Rio, Martin. *Disquisitionum magicarum libri sex*. Ursel, 1606.

————. *Investigations into Magic*. Translated and edited by P. G. Maxwell-Stuart. Manchester: Manchester University Press, 2000.

Demaitre, Luke M. *Doctor Bernard de Gordon: Professor and Practitioner*. Toronto: Pontifical Institute of Medieval Studies, 1980.

Deutsches Theatrum Chemicum. Edited by Friedrich Roth-Scholtz. 3 vols. Nuremberg, 1728.

DeVun, Leah. "The Jesus Hermaphrodite: Science and Sex Difference in Premodern Europe." *Journal for the History of Ideas* 69 (2008): 193–218.

————. *Prophecy, Alchemy, and the End of Time: John of Rupescissa in the Late Middle Ages*. New York: Columbia University Press, 2009.

Digby, Kenelm. *A Choice Collection of Rare Chymical Secrets*. London, 1682.

————. *A Discourse on the Vegetation of Plants*. London, 1661.

Dobbs, Betty Jo Teeter. *The Foundations of Newton's Alchemy; or, Hunting of the Greene Lyon*. Cambridge: Cambridge University Press, 1975.

————. *The Janus Faces of Genius*. Cambridge: Cambridge University Press, 1991.

————. "Newton's Commentary on *The Emerald Tablet* of Hermes Trismegestus: Its Scientific and Theological Significance." In *Hermeticism and the Renaissance*, edited by Ingrid Merkel and Allen G. Debus, pp. 182–91. Washington, DC: Folger Shakespeare Library, 1988.

————. "Studies in the Natural Philosophy of Sir Kenelm Digby: Part I." *Ambix* 18 (1971): 1–25.

————. "Studies in the Natural Philosophy of Sir Kenelm Digby: Part II." *Ambix* 20 (1973): 143–63.

————. "Studies in the Natural Philosophy of Sir Kenelm Digby: Part III." *Ambix* 21 (1974): 1–28.

Dorn, Gerhard. *Physica Trismegesti*. In *Theatrum chemicum*, 1:362–87.

Duchesne, Joseph. *Ad veritatem Hermeticae medicinae*. Paris, 1604.

Duclo, Gaston. *De triplici praeparatione argenti et auri*. In *Theatrum chemicum*, 4:371–88.

Duncan, Edgar H. "The Alchemy in Jonson's *Mercury Vindicated*." *Studies in Philology* 39 (1942): 625–37.

————. "Donne's Alchemical Figures." *English Literary History* 9 (1942): 257–85.

————. "Jonson's Alchemist and the Literature of Alchemy." *Proceedings of the Modern Language Association* 61 (1946): 699–710.

————. "The Literature of Alchemy and Chaucer's Canon's Yeoman's Tale: Framework, Theme, and Characters." *Speculum* 43 (1968): 633–56.

————. "The Yeoman's Canon's 'Silver Citrinacioun.'" *Modern Philology* 37 (1940): 241–62.

Durocher, Alain, and Antoine Faivre, eds. *Die templerische und okkultistische Freimaurerei im 18. und 19. Jahrhundert*. 4 vols. Leimen, Germany: Kristkeitz, 1987–92.

Duveen, Denis. "James Price (1752–1783) Chemist and Alchemist." *Isis* 41 (1950): 281–83.

Eamon, William. "Alchemy in Popular Culture: Leonardo Fioravanti and the Search for the Philosopher's Stone." *Early Science and Medicine* 5 (2000): 196–213.

Eliade, Mircea. *The Forge and the Crucible*. Chicago: University of Chicago Press, 1978. Originally published as *Forgerons et alchimistes*. Paris: Flammarion, 1956.

————. "Metallurgy, Magic and Alchemy." Cahiers de Zalmoxis, 1. Paris: Librairie Orientaliste Paul Geuthner, 1938.

Emerson, Jocelyn. "John Donne and the Noble Art." In *Textual Healing: Essays in Medieval and Early Modern Medicine*, edited by Elizabeth Lane Furdell, pp. 195–221. Leiden: Brill, 2005.

Eymerich, Nicolas. *Contra alchemistas*. Edited by Sylvain Matton. *Chrysopoeia* 1 (1987): 93–136.

Fabre, François. "Robert Fludd et l'alchimie: Le *Tractatus Apologeticus integritatem societatis de Rosea Cruce defendens*." *Chrysopoeia* 7 (2000–2003): 251–91.

Fabre, Pierre-Jean. *Alchymista christianus*. Toulouse, 1632. Reprinted, with accompanying French translation, as *L'alchimiste chrétien*, edited and translated by Frank Grenier. Textes et Travaux de Chrysopoeia, 7. Paris: SÉHA; Milan: Archè, 2001.

———. *Hercules piochymicus*. Toulouse, 1634.

Fabrizio-Costa, Sylvia. "De quelques emplois des thémes alchimiques dans l'art oratoire italien du XVIIᵉ siècle." *Chrysopoeia* 3 (1989): 135–62.

Faivre, Antoine, ed. *René Le Forestier: La Franc-Maçonnerie templiére et occultiste aux XVIIIᵉ et XIXᵉ siècles*. Paris: Aubier-Montaigne, 1970.

Fanianus, Johannes Chrysippus. *De jure artis alchimiae*. In *Theatrum chemicum*, 1:48–63.

Festugiére, A. J. *La révélation d'Hermés Trismégeste*. Paris: Librarie Lecoffre, 1950.

Figala, Karin, and Ulrich Neumann. "'Author, Cui Nomen Hermes Malavici': New Light on the Biobibliography of Michael Maier (1569–1622)." In Rattansi and Clericuzio, *Alchemy and Chemistry in the Sixteenth and Seventeenth Centuries*, pp. 121–48.

———. "À propos de Michel Maier: Quelques découvertes bio-bibliographiques." In Kahn and Matton, *Alchimie: Art, histoire, et mythes*, pp. 651–64.

Figuier, Louis. *L'Alchimie et les alchimistes*. 2nd ed. Paris, 1856.

Fischer, Hermann. *Metaphysische, experimentelle und utilitaristische Traditionen in der Antimonliteratur zur Zeit der "wissenschaftlichen Revolution": Eine kommentierte Auswahl-Bibliographie*. Braunschweiger Veröffenlichungen zu Geschichte der Pharmazie und der Naturwissenschaften. Brunswick, 1988.

Flamel, Nicolas, pseudo-. *Exposition of the Hieroglyphicall Figures*. London, 1624. Reprint, New York: Garland, 1994.

Forshaw, Peter J. "Alchemy in the Amphitheatre: Some Considerations of the Alchemical Content of the Engravings in Heinrich Khunrath's *Amphitheatre of Eternal Wisdom* (1609)." In Wamberg, *Alchemy and Art*, pp. 195–220.

———. "Vitriolic Reactions: Orthodox Responses to the Alchemical Exegesis of Genesis." In *The Word and the World: Biblical Exegesis and Early Modern Science*, edited by Kevin Killeen and Peter J. Forshaw, pp. 111–36. Basingstoke: Palgrave, 2007.

Franck de Franckenau, Georg, and Johann Christian Nehring. *De Palingenesia*. Halle, 1717.

Fück, J. W. "The Arabic Literature on Alchemy according to An-Nadīm." *Ambix* 4 (1951): 81–144.

Ganzenmüller, Wilhelm. "Das Buch der heiligen Dreifaltigkeit." *Archiv der Kulturgeschichte* 29 (1939): 93–141.

Garber, Margaret. "Transitioning from Transubstantiation to Transmutation: Catholic Anxieties over Chymical Matter Theory at the University of Prague." In Principe, *Chymists and Chymistry*, pp. 63–76.

Garbers, Karl, and Jost Weyer, eds. *Quellengeschichtliches Lesebuch zur Chemie und Alchemie der Araber im Mittelalter*. Hamburg: Helmut Buske Verlag, 1980.

Ge, Hong. *Alchemy, Medicine, Religion in the China of AD 320.* Cambridge, MA: MIT Press, 1967.

Geffarth, Renko. *Religion und arkane Hierarchie: Der Orden der Gold- und Rosenkreuzer als geheime Kirche im 18. Jahrhundert.* Leiden: Brill, 2007.

Das Geheimnis aller Geheimnisse . . . oder der güldene Begriff der geheimsten Geheimnisse der Rosen- und Gülden-Kreutzer. Leipzig, 1788.

Geoffroy, Étienne-François. "Des supercheries concernant la pierre philosophale." *Mémoires de l'Académie Royale des Sciences* 24 (1722): 61–70.

Geoghegan, D. "A Licence of Henry VI to Practise Alchemy." *Ambix* 6 (1957):10–17.

Gilbert, R. A. *A. E. Waite: Magician of Many Parts.* Wellingborough, UK: Crucible, 1987.

———. *The Golden Dawn: Twilight of the Magicians.* San Bernardino, CA: Borgo Press, 1988.

Gmelins Handbuch der anorganischen Chemie. Leipzig: Verlag Chemie, 1924–.

Goltz, Dietlinde. "Alchemie und Aufklärung: Ein Beitrag zur Naturwissenschafts-geschichtsschreibung der Aufklärung." *Medizinhistorische Journal* 7 (1972): 31–48.

Grafton, Anthony. "Protestant versus Prophet: Isaac Casaubon on Hermes Trismegistus." *Journal of the Warburg and Courtauld Institutes* 46 (1983): 78–93.

Grant, Edward. *The Foundations of Modern Science in the Middle Ages.* Cambridge: Cambridge University Press, 1996.

Grell, Ole Peter, ed. *Paracelsus: The Man and His Reputation, His Ideas and Their Transformation.* Leiden: Brill, 1998.

Gruman, Gerald J. *A History of Ideas about the Prolongation of Life.* Philadelphia: American Philosophical Society, 1966. Reprint, New York: Arno Press, 1977.

Guerrero, José Rodríguez. "Some Forgotten Fez Alchemists and the Loss of the Peñon de Vélez de la Gomera in the Sixteenth Century." In *Chymia: Science and Nature in Medieval and Early Modern Europe,* edited by Miguel López-Pérez, Didier Kahn, and Mar Rey Bueno, pp. 291–309. Newcastle-upon-Tyne: Cambridge Scholars, 2010.

Güldenfalk, Siegmund Heinrich. *Sammlung von mehr als hundert wahrhaften Transmu-tationgeschichten.* Frankfurt, 1784.

Gutas, Dimitri. *Greek Thought, Arabic Culture: The Graeco-Arabic Translation Movement in Baghdad and Early 'Abbasid Society.* London: Routledge, 1998.

Halleux, Robert. "Albert le Grand et l'alchimie." *Revue des sciences philosophiques et théologiques* 66 (1982): 57–80.

———. "L'alchimiste et l'essayeur." In Meinel, *Die Alchemie in der europäischen Kultur- und Wissenschaftsgeschichte,* pp. 277–91.

———. *Les alchimistes grecs I: Papyrus de Leyde, Papyrus de Stockholm, Recettes.* Paris: Les Belles Lettres, 1981.

———. "La controverse sur les origines de la chimie de Paracelse à Borrichius." In *Acta conventus neo-latini Turonensis,* 2:807–17. Paris: Vrin, 1980.

———. "Le mythe de Nicolas Flamel, ou les mécanismes de la pseudépigraphie alchi-mique." *Archives internationales de l'histoire des sciences* 33 (1983): 234–55.

———. "Ouvrages alchimiques de Jean de Rupescissa." In *Histoire littéraire de la France,* 41:241–77.

———. *Le problème des métaux dans la science antique.* Paris: Les Belles Lettres, 1974.

———. "La réception de l'alchimie arabe en Occident." In Rashed and Morelon, *Histoire des sciences arabes,* 3:143–54.

————. *Les textes alchimiques.* Turnhout, Belgium: Brepols, 1979.

————. "Theory and Experiment in the Early Writings of Johan Baptist Van Helmont." In *Theory and Experiment*, edited by Diderik Batens, pp. 93–101. Dordrecht: Rediel, 1988.

Hallum, Benjamin C. Essay review of the *Tome of Images. Ambix* 56 (2009): 76–88.

————. "Zosimus Arabus." PhD diss., Warburg Institute, 2008.

Hanegraaff, Wouter J. *Esotericism and the Academy: Rejected Knowledge in Western Culture.* Cambridge: Cambridge University Press, 2012.

Hanegraaff, Wouter J., Antoine Faivre, Roelof van den Broek, and Jean-Pierre Brach, eds. *The Dictionary of Gnosis and Western Esotericism.* 2 vols. Leiden: Brill, 2005.

Hannaway, Owen. "Georgius Agricola as Humanist." *Journal of the History of Ideas* 53 (1992): 553–60.

Harkness, Deborah. *John Dee's Conversations with Angels: Cabala, Alchemy, and the End of Nature.* Cambridge: Cambridge University Press, 1999.

Hartog, P. J., and E. L. Scott. "Price, James (1757/8–1783)." *Oxford Dictionary of National Biography*, s.v. Oxford: Oxford University Press, 2004.

Haskins, Charles Homer. *The Renaissance of the Twelfth Century.* Cambridge, MA: Harvard University Press, 1927.

Hassan, Ahmad Y. "The Arabic Original of the *Liber de compositione alchemiae.*" *Arabic Sciences and Philosophy* 14 (2004): 213–31.

Helvetius, Johann Friedrich. *Vitulus aureus.* In *Musaeum hermeticum*, pp. 815–63.

Hirai, Hiro. *Le concept de semence dans les théories de la matière à la Renaissance de Marsile Ficin à Pierre Gassendi.* Turnhout, Belgium: Brepols, 2005.

Histoire litteraire de la France. 41 vols. Paris: Academie des Insciptions et Belles-Lettres, 1981.

Hitchcock, Ethan Allen. *Remarks upon Alchemy and the Alchemists.* Boston, 1857. Reprint, New York: Arno Press, 1976.

————. *Remarks upon Alchymists.* Carlisle, PA, 1855.

Hoffmann, Klaus. *Johann Friedrich Böttger: Vom Alchemistengold zum weissen Porzellan.* Berlin: Verlag Neues Leben, 1985.

Hoghelande, Ewald van. *Historiae aliquot transmutationis metallicae . . . pro defensione alchymiae contra hostium rabiem.* Cologne, 1604.

Holmyard, E. J. *Alchemy.* Harmondsworth: Penguin, 1957.

————, ed. and trans. *The Arabic Works of Jābir ibn Hayyān.* Paris: Geuthner, 1928.

————. "The Emerald Table." *Nature* 112 (1923): 525–26.

————. "Jābir ibn-Hayyān." *Proceedings of the Royal Society of Medicine, Section of the History of Medicine* 16 (1923): 46–57.

Howe, Ellic, ed. *The Alchemist of the Golden Dawn: The Letters of the Reverend W. A. Ayton to F. L. Gardner and Others 1886–1905.* Wellingborough, UK: Aquarian Press, 1985.

————. *The Magicians of the Golden Dawn.* New York: Samuel Weiser, 1978.

Hudry, Françoise, ed. "Le *De secretis naturae* du pseudo-Apollonius de Tyane: Traduction latine par Hugues de Santalla du *Kitāb sirr al-ḫalīqa* de Balīnūs." In "Cinq traités alchimique médiévaux," *Chrysopoeia* 6 (1997–99): 1–153.

Hunter, Michael. "Alchemy, Magic, and Moralism in the Thought of Robert Boyle." *British Journal for the History of Science* 23 (1990): 387–410.

————. *Robert Boyle by Himself and His Friends.* London: Pickering, 1994.

Hunter, Michael, and Lawrence M. Principe. "The Lost Papers of Robert Boyle." *Annals of Science* 60 (2003): 269–311.

Husson, Bernard. *Transmutations alchimiques*. Paris: Editions J'ai Lu, 1974.

Ibn-Khaldūn. *The Muqaddimah: An Introduction to History*. 3 vols. New York: Pantheon, 1958.

Ibn-Sīnā. *Avicennae de congelatione et conglutinatione lapidum, Being Sections of the Kitāb al-Shifā'*. Edited by E. J. Holmyard and D. C. Mandeville. Paris: Paul Geuthner, 1927.

———. *Avicennae de congelatione et conglutinatione lapidum*. In *Bibliotheca chemica curiosa*, 1:636–38.

Jābir ibn-Hayyān. *Das Buch der Gifte*. Edited by Alfred Siggel. Wiesbaden: Akademie der Wissenschaften und der Literatur, 1958.

———. *Dix traités d'alchimie*. Translated by Pierre Lory. Paris: Sinbad, 1983.

———. "Liber Misericordiae Geber: Eine lateinische Übersetzung des grösseren Kitāb alrahma." Edited by Ernst Darmstaedter. *Archiv für Geschichte der Medizin* 17 (1925): 187–97.

Jantz, Harold. "Goethe, Faust, Alchemy, and Jung." *German Quarterly* 35 (1962): 129–41.

Jennings, Hargrave. *The Rosicrucians*. London, 1870.

John of Antioch. *Iohannes Antiocheni fragmenta ex Historia chronica*. Edited and translated by Umberto Roberto. Berlin: De Gruyter, 2005.

John of Rupescissa. *The Book of the Quinte Essence*. Edited by F. J. Furnivall. London: Early English Text Society, 1866. Reprint, Oxford: Oxford University Press, 1965.

———. *De confectione veri lapidis philosophorum*. In *Bibliotheca chemica curiosa*, 2:80–83.

———. *Liber lucis*. In *Bibliotheca chemica curiosa*, 2:84–87.

Johnson, Rozelle Parker. *Compositiones variae: An Introductory Study*. Illinois Studies in Language and Literature 23. Urbana, 1939.

Jollivet-Castelot, François. *Comment on devient alchimiste*. Paris, 1897.

———. *La révolution chimique et la transmutation des métaux*. Paris: Chacornac, 1925.

———. *La synthèse de l'or*. Paris: Daragon, 1909.

———. *Synthèse des sciences occultes*. Paris, 1928.

Joly, Bernard. "L'alkahest, dissolvant universel, ou quand la thèorie rend pensible une pratique impossible." *Revue d'histoire des sciences* 49 (1996): 308–30.

———. *La rationalité de l'alchimie au XVIIᵉ siècle*. Paris: Vrin, 1992.

———. "La rationalité de l'Hermétisme: La figure d'Hermès dans l'alchimie à l'âge classique." *Methodos* 3 (2003): 61–82.

Jong, H. M. E. de. *Michael Maier's Atalanta Fugiens: Sources of an Alchemical Book of Emblems*. Leiden: Brill, 1969.

Jung, Carl Gustav. *Collected Works of Carl Gustav Jung* (20 vols.): vol. 9, pt. 2: *Aion*; vol. 12: *Psychology and Alchemy*; vol. 13: *Alchemical Studies*; vol. 14: *Mysterium Conjunctionis*. London: Routledge, 1953–79.

———. "Die Erlösungsvorstellungen in der Alchemie." *Eranos-Jahrbuch 1936*. Zurich: Rhein-Verlag, 1937.

———. "The Idea of Redemption in Alchemy." In *The Integration of the Personality*, edited by Stanley Dell, pp. 205–80. New York: Farrar and Rinehart, 1939.

Kahn, Didier. "Alchemical Poetry in Medieval and Early Modern Europe: A Preliminary Survey and Synthesis; Part I: Preliminary Survey." *Ambix* 57 (2010): 249–74.

————. *Alchimie et Paracelsianisme en France (1567–1625)*. Geneva: Droz, 2007.

————. "Les débuts de Gérard Dorn." In *Analecta Paracelsica: Studien zum Nachleben Theophrast von Hohenheims im deutschen Kulturgebiet der frühen Neuzeit*, edited by Joachim Telle, pp. 59–126. Stuttgart: Franz Steiner Verlag, 1994.

————. "L'interprétation alchimique de la Genèse chez Joseph Du Chesne dans le contexte de ses doctrines alchimiques et cosmologiques." In *Scientiae et artes: Die Vermittlung alten und neuen Wissens in Literatur, Kunst und Musik*, edited by Barbara Mahlmann-Bauer, pp. 641–92. Wiesbaden, Harrassowitz, 2004.

————. "King Henry IV, Alchemy, and Paracelsianism in France (1589–1610)." In *Principe, Chymists and Chymistry*, pp. 1–11.

————, ed. *La table d'émeraude et sa tradition alchimique*. Paris: Belles Lettres, 1994.

Kahn, Didier, and Sylvain Matton, eds. *Alchimie: Art, histoire, et mythes*. Textes et Travaux de Chrysopoeia 1. Paris: SÉHA; Milan: Archè, 1995.

Kane, Robert. *Elements of Chemistry*. New York, 1842.

Karpenko, Vladimir. *Alchemical Coins and Medals*. Glasgow: Adam Maclean, 1998.

————. "Alchemistische Münzen und Medaillen." *Anzeiger der Germanisches National-museums 2001*, pp. 49–72. Nuremberg: Germanisches Nationalmuseum, 2001.

————. "Coins and Medals Made of Alchemical Metal." *Ambix* 35 (1988): 65–76.

————. "Systems of Metals in Alchemy." *Ambix* 50 (2003): 208–30.

Kauffman, George B. "The Mystery of Stephen H. Emmens: Successful Alchemist or Ingenious Swindler?" *Ambix* 30 (1983): 65–88.

Keyser, Paul T. "Greco-Roman Alchemy and Coins of Imitation Silver." *American Journal of Numismatics* 7–8 (1995): 209–33.

Khunrath, Heinrich. *Lux in tenebris*. N.p., 1614.

————. *Trewhertzige Warnungs-Vermahnung*. Magdeburg, 1597.

Kibre, Pearl. "Albertus Magnus on Alchemy." In *Albertus Magnus and the Sciences: Commemorative Essays 1980*, edited by James A. Weisheipl, pp. 187–202. Toronto: Pontifical Institute of Mediaeval Studies, 1980.

————. "Alchemical Writings Attributed to Albertus Magnus." *Speculum* 17 (1942): 511–15.

Kircher, Athanasius. *Mundus subterraneus*. Amsterdam, 1678.

Klein-Francke, Felix. "Al-Kindi." In *The History of Islamic Philosophy*, edited by Seyyed Hossein Nasr and Oliver Leaman, pp. 165–77. New York: Routledge, 1996.

Kraus, Paul. *Jābir ibn Hayyān: Contribution à l'histoire des idées scientifiques dans l'Islam*. Vol. 1, *Le Corpus des écrits jābiriens*. Mémoires de L'Institut d'Égypte 44 (1943).

————. *Jābir ibn Hayyān: Contribution à l'histoire des idées scientifiques dans l'Islam*. Vol. 2, *Jābir et la science grecque*. Mémoires de L'Institut d'Égypte 45 (1942). Reprint, Paris: Les Belles Lettres, 1986.

————, ed. *Jābir ibn-Hayyān: Textes choisis*. Paris: Maisonneuve, 1935.

Lambsprinck. *De lapide philosophico*. In *Musaeum hermeticum*, pp. 337–71.

Lapidus. *In Pursuit of Gold: Alchemy in Theory and Practice*. New York: Samuel Weiser, 1976.

Leibenguth, Erik. *Hermetische Poesie des Frühbarock: Die "Cantilenae intellectuales" Michael Maiers*. Tübingen: Max Niemeyer Verlag, 2002.

Leibniz, Gottfried Wilhelm. "Oedipus chymicus." *Miscellanea Berolinensia* 1 (1710): 16–21.

Lemay, Richard. "L'authenticité de la Préface de Robert de Chester à sa traduction du *Morienus*." *Chrysopoeia* 4 (1990–91): 3–32.

Lemery, Nicolas. *Cours de chymie.* Paris: 1683.

Lenglet du Fresnoy, Nicolas. *Histoire de la philosophie hermetique.* 3 vols. Paris, 1742–44.

Lennep, Jacques van. *Art et alchimie.* Brussels: Meddens, 1966.

Lenz, Hans Gerhard, ed. *Triumphwagen des Antimons: Basilius Valentinus, Kerckring, Kirchweger; Text, Kommentare, Studien.* Elberfeld, Germany: Humberg, 2004.

Leo Africanus. *A Geographicall Historie of Africa.* London, 1600.

Le Pelletier, Jean. *L'Alkaest; ou, Le dissolvant universel de Van Helmont.* Rouen, 1706.

Levey, Martin. *Chemistry and Chemical Technologies in Ancient Mesopotamia.* Amsterdam: Elsevier, 1959.

Lindberg, David C. *The Beginnings of Western Science.* 2nd ed. Chicago: University of Chicago Press, 2007.

Linden, Stanton J. *Darke Hieroglyphicks: Alchemy in English Literature from Chaucer to the Restoration.* Lexington: University Press of Kentucky, 1996.

———. "Jonson and Sendivogius: Some New Light on 'Mercury Vindicated.'" *Ambix* 24 (1977): 39–54.

Lloyd, G. E. R. *Greek Science after Aristotle.* New York: Norton, 1973.

López-Pérez, Miguel, Didier Kahn, and Mar Rey Bueno, eds. *Chymia: Science and Nature in Medieval and Early Modern Europe.* Newcastle-upon-Tyne: Cambridge Scholars Publishing, 2010.

Lory, Pierre, ed. *L'Élaboration de l'Élixir Suprême.* Damascus: Institut Français de Damas, 1988.

Luca, Alfred, and John R. Harris. *Ancient Egyptian Materials and Industries.* London: Arnold, 1962.

Lüthy, Christoph. "The Fourfold Democritus on the Stage of Early Modern Europe." *Isis* 91 (2000): 442–79.

Magdalino, Paul, and Maria Mavroudi. *The Occult Sciences in Byzantium.* Geneva: La Pomme d'Or, 2006.

Maier, Michael. *Arcana arcanissima.* London, 1613.

———. *Atalanta fugiens.* Oppenheim, Germany, 1618.

———. *Examen fucorum pseudo-chymicorum detectorum et in gratiam veritatis amantium succincte refutatorum.* Frankfurt, 1617.

———. *Tripus aureus.* Frankfurt, 1618.

Malcolm, Noel. "Robert Boyle, Georges Pierre des Clozets, and the Asterism: New Sources." *Early Science and Medicine* 9 (2004): 293–306.

Mandosio, Jean-Marc. "La place de l'alchimie dans les classifications des sciences et des arts à la Renaissance." *Chrysopoeia* 4 (1990–91): 199–282.

Margolin, Jean-Claude, and Sylvain Matton, eds. *Alchimie et philosophie à la Renaissance.* Paris: Vrin, 1993.

Martelli, Matteo. "Chymica Graeco-Syriaca: Osservationi sugli scritti alchemici pseudo-Democritei nelle tradizioni greca e sirica." In *'Uyūn al-Akhbār: Studi sul mondo Islamico; Incontro con l'altro e incroci di culture,* edited by D. Cevenini and S. D'Onofrio, pp. 219–49. Bologna: Il Ponte, 2008.

———. "'Divine Water' in the Alchemical Writings of Pseudo-Democritus." *Ambix* 56 (2009): 5–22.

———. "Greek Alchemists at Work: 'Alchemical Laboratory' in the Greco-Roman Egypt." *Nuncius* 26 (2011): 271–311.

———. "L'opera alchemica dello Pseudo-Democrito: Un riesame del testo." *Eikasmos* 14 (2003): 161–84.

———, ed. *Pseudo-Democrito: Scritti alchemici, con il commentario di Sinesio; Edizione critica del testo greco, traduzione e commento.* Textes et Travaux de Chrysopoeia 12. Paris: SÉHA; Milan: Archè, 2011.

Martels, Zweder van. "Augurello's *Chrysopoeia* (1515): A Turning Point in the Literary Tradition of Alchemical Texts." *Early Science and Medicine* 5 (2000): 178–95.

Martin, Craig. "Alchemy and the Renaissance Commentary Tradition on *Meteorologica* IV." *Ambix* 51 (2004): 245–62.

Martin, Luther H. "A History of the Psychological Interpretation of Alchemy." *Ambix* 22 (1975): 10–20.

Martinez Oliva, Juan Carlos. "Monetary Integration in the Roman Empire." In *From the Athenian Tetradrachm to the Euro*, edited by P. L. Cottrell, Gérasimos Notaras, and Gabriel Tortella, pp. 7–23. Burlington, VT: Ashgate, 2007.

Martinón-Torres, Marcos. "Some Recent Developments in the Historiography of Alchemy." *Ambix* 58 (2011): 215–37.

Martinón-Torres, Marcos, and Thilo Rehren. "Alchemy, Chemistry and Metallurgy in Renaissance Europe: A Wider Context for Fire Assay Remains." *Historical Metallurgy* 39 (2005): 14–31.

———. "Post-Medieval Crucible Production and Distribution: A Study of Materials and Materialities." *Archaeometry* 51 (2009): 49–74.

Martinón-Torres, Marcos, Thilo Rehren, and I. C. Freestone. "Mullite and the Mystery of Hessian Wares." *Nature* 444 (2006): 437–38.

Marx, Jacques. "Alchimie et Palingénésie." *Isis* 62 (1971): 274–89.

Matton, Sylvain. "L'influence de l'humanisme sur la tradition alchimique." In "Le crisi dell'alchemia," *Micrologus* 3 (1995): 279–345.

———. "L'interprétation alchimique de la mythologie." *Dix-huitième siècle* 27 (1995): 73–87.

———. "Une lecture alchimique de la Bible: Les 'Paradoxes chimiques' de Francois Thybourel." *Chrysopoeia* 2 (1988): 401–22.

———. "Remarques sur l'alchimie transmutatoire chez les théologiens réformés de la Renaissance." *Chrysopoeia* 7 (2000–2003): 171–87.

———. *Scolastique et alchimie.* Textes et Travaux de Chrysopoeia 10. Paris: SÉHA; Milan: Archè, 2009.

———. "Thématique alchimique et litterature religieuse dans la France du XVIIᵉ siècle." *Chrysopoeia* 2 (1998): 129–208.

McGuire, J. E., and P. M. Rattansi. "Newton and the Pipes of Pan." *Notes and Records of the Royal Society of London* 21 (1966): 108–43.

McIntosh, Christopher. *Eliphas Lévi and the French Occult Revival.* London: Rider, 1975.

———. *The Rose Cross and the Age of Reason: Eighteenth Century Rosicrucianism in Central Europe and Its Relationship to the Enlightenment.* Leiden: Brill, 1992.

Mehrens, A. F. "Vues d'Avicenne sur astrologie et sur le rapport de la responsabilité humaine avec le destin." *Muséon* 3 (1884): 383–403.

Meinel, Christoph. "Alchemie und Musik." In *Die Alchemie in der europäischer Kultur- und Wissenschaftsgeschichte*, pp. 201–28.

———, ed. *Die Alchemie in der europäischer Kultur- und Wissenschaftsgeschichte.* Wölfenbütteler Forschungen 32. Wiesbaden: Harrassowitz, 1986.

Mellor, J. W. *A Comprehensive Treatise on Inorganic and Theoretical Chemistry.* 16 vols. London: Longmans, 1922–37.

Mercier, Alain. "August Strindberg et les alchimistes français: Hemel, Vial, Tiffereau, Jollivet-Castelot." *Revue de littérature comparée* 43 (1969): 23–46.

Merkur, Dan. "Methodology and the Study of Western Spiritual Alchemy." *Theosophical History* 8 (2000): 53–70.

Mertens, Michèle. *Les alchimistes grecs IV, i: Zosime de Panopolis, Mémoires authentiques.* Paris: Les Belles Lettres, 2002.

———. "Graeco-Egyptian Alchemy in Byzantium." In Magdalino and Mavroudi, *The Occult Sciences in Byzantium*, pp. 205–30.

Minnen, Peter van. "Urban Craftsmen in Roman Egypt." *Münstersche Beiträge zur antiken Handelsgeschichte* 6 (1987): 31–87.

Möller, H. "Die Gold- und Rosenkreuzer, Struktur, Zielsetzung und Wirkung einer anti-aufklärerischen Geheimgesellschaft." In *Geheime Gesellschaften*, edited by Peter Christian Ludz, pp. 153–202. Heidelberg: Schneider, 1979.

Moran, Bruce T. *The Alchemical World of the German Court.* Sudhoffs Archiv 29. Stuttgart: Franz Steiner Verlag, 1991.

———. "Alchemy and the History of Science: Introduction." *Isis* 102 (2011): 300–304.

———. *Andreas Libavius and the Transformation of Alchemy: Separating Chemical Cultures with Polemical Fire.* Sagamore Beach, MA: Science History Publications, 2007.

———. *Distilling Knowledge: Alchemy, Chemistry, and the Scientific Revolution.* Cambridge, MA: Harvard University Press, 2005.

Morhof, Daniel Georg. *De metallorum transmutatione epistola.* In Manget, *Bibliotheca chemica curiosa*, 1:168–92.

Morienus. *De compositione alchemiae.* In *Bibliotheca chemica curiosa*, 1:509–19.

———. *A Testament of Alchemy.* Edited and translated by Lee Stavenhagen. Hanover, NH: University Press of New England / Brandeis University Press, 1974.

Musaeum hermeticum. Frankfurt, 1678. Reprint, Graz: Akademische Druck, 1970.

Muslow, Martin. "Ambiguities of the *Prisca Sapientia* in Late Renaissance Humanism." *Journal of the History of Ideas* 65 (2004): 1–13.

Needham, Joseph. "The Elixir Concept and Chemical Medicine in East and West." *Organon* 11 (1975): 167–92.

———. *Science and Civilisation in China.* Vol. 5, *Chemistry and Chemical Technology.* Cambridge: Cambridge University Press, 1974–83.

Neumann, Ulrich. "Michel Maier (1569–1622): 'Philosophe et médecin.'" In Margolin and Matton, *Alchimie et philosophie à la Renaissance*, pp. 307–26.

Newman, William R. *Atoms and Alchemy.* Chicago: University of Chicago Press, 2006.

———. *Gehennical Fire: The Lives of George Starkey, an American Alchemist in the Scientific Revolution.* Cambridge, MA: Harvard University Press, 1994.

———. "Genesis of the *Summa perfectionis*." *Archives internationales d'histoire des sciences* 35 (1985): 240–302.

———. "The Homunculus and His Forebears: Wonders of Art and Nature." In *Natural Particulars: Nature and the Disciplines in Renaissance Europe*, edited by Anthony Grafton and Nancy Siraisi, pp. 321–45. Cambridge, MA: MIT Press, 1999.

———. "New Light on the Identity of Geber." *Sudhoffs Archiv* 69 (1985): 79–90.

———. "Newton's *Clavis* as Starkey's *Key*." *Isis* 78 (1987): 564–74.

———. "The Philosophers' Egg: Theory and Practice in the Alchemy of Roger Bacon." In "Le crisi dell'alchimia," *Micrologus* 3 (1995): 75–101.

———. *Promethean Ambitions: Alchemy and the Quest to Perfect Nature.* Chicago: University of Chicago Press, 2004.

———. *The Summa Perfectionis of the Pseudo-Geber: A Critical Edition, Translation, and Study*. Leiden: Brill, 1991.

———. "Technology and Alchemical Debate in the Late Middle Ages." *Isis* 80 (1989): 423–45.

Newman, William R., and Lawrence M. Principe. *Alchemy Tried in the Fire: Starkey, Boyle, and the Fate of Helmontian Chymistry*. Chicago: University of Chicago Press, 2002.

———. "Alchemy vs. Chemistry: The Etymological Origins of a Historiographic Mistake." *Early Science and Medicine* 3 (1998): 32–65.

Nicholson, Paul T., and Ian Shaw, eds. *Ancient Egyptian Materials and Technology*. Cambridge: Cambridge University Press, 2000.

Noll, Richard. *The Aryan Christ*. New York: Random House, 1997.

———. *The Jung Cult*. Princeton, NJ: Princeton University Press, 1994.

Norton, Thomas. *Ordinall of Alchimy*. In Ashmole, *Theatrum chemicum britannicum*.

Nummedal, Tara. *Alchemy and Authority in the Holy Roman Empire*. Chicago: University of Chicago Press, 2007.

———. "Words and Works in the History of Alchemy." *Isis* 102 (2011): 330–37.

Obrist, Barbara. *Les débuts de l'imagerie alchimique*. Paris: Le Sycomore, 1982.

Opsomer, Carmélia, and Robert Halleux. "L'Alchimie de Théophile et l'abbaye de Stavelot." In *Comprendre et maîtriser la nature au Moyen Age*, edited by Guy Beaujouan, pp. 437–59. Geneva: Droz, 1994.

Osler, Margaret J. *Reconfiguring the World: Nature, God, and Human Understanding from the Middle Ages to Early Modern Europe*. Baltimore: Johns Hopkins University Press, 2010.

Pagel, Walter. *Joan Baptista Van Helmont*. Cambridge: Cambridge University Press, 1982.

———. *Paracelsus: An Introduction to Philosophical Medicine in the Era of the Renaissance*. Basel: Karger, 1958.

Paneth, Fritz. "Ancient and Modern Alchemy." *Science* 64 (1926): 409–17.

Pantheus. *Voarchadumia*. In *Theatrum chemicum*, 2:495–549.

Papathanassiou, Maria K. "L'Oeuvre alchimique de Stephanos d'Alexandrie." In Viano, *L'Alchimie et ses racines philosophiques*, pp. 113–33.

———. "Stephanos of Alexandria: A Famous Byzantine Scholar, Alchemist and Astrologer." In Magdalino and Mavroudi, *The Occult Sciences in Byzantium*, pp. 163–203.

———. "Stephanus of Alexandria: On the Structure and Date of His Alchemical Work." *Medicina nei secoli* 8 (1996): 247–66.

Paracelsus, [pseudo?]. *De rerum natura*. In *Sämtliche Werke*, edited by Karl Sudoff. *Abteilung 1: Medizinische, wissenschaftliche, und philosophische Schriften* (Munich: Oldenbourg, 1922–33), 11:316–17.

Percolla, Vincenzo. *Auriloquio*. Edited by Carlo Alberto Anzuini. Textes et Travaux de Chrysopoeia 2. Paris: SÉHA; Milan: Archè, 1996.

Pereira, Michela. *The Alchemical Corpus Attributed to Raymond Lull*. London: Warburg Institute, 1989.

———. "La leggenda di Lullo alchimista." *Estudios lulianos* 27 (1987): 145–63.

———. "*Medicina* in the Alchemical Writings Attributed to Raimond Lull." In Rattansi and Clericuzio, *Alchemy and Chemistry in the Sixteenth and Seventeenth Centuries*, pp. 1–15.

————. "Sulla tradizione testuale del *Liber de secretis naturae seu de quinta essentia* attribuito a Raimondo Lullo." *Archives internationales d'histoire des sciences* 36 (1986): 1–16.

————. "Teorie dell'elixir nell'alchimia latina medievale." In "Le crisi dell'alchimia," *Micrologus* 3 (1995): 103–48.

————. "Un tesoro inestimabile: Elixir e *prolongatio vitae* nell'alchimiae del'300." *Micrologus* 1 (1992): 161–87.

Pereira, Michela, and Barbara Spaggiari. *Il Testamentum alchemico attribuito a Raimondo Lullo*. Florence: Sismel, 1999.

Perifano, Alfredo. "Theorica et practica dans un manuscrit alchimique de Sisto de Boni Sexti da Norcia, alchimiste à la cour de Côme Ier de Médicis." *Chrysopoeia* 4 (1990–91): 81–146.

Pernety, Antoine-Joseph. *Dictionnaire mytho-hermétique*. Paris, 1758.

————. *Les fables égyptiennes et grecques dévoilées*. 2 vols. Paris, 1758. Reprint, Paris: La Table d'émeraude, 1982.

Petrarch. *Remedies for Fortune Fair and Foul*. Translated by Conrad H. Rawski. 5 vols. Bloomington: Indiana University Press, 1991.

Petrus Bonus. *Margarita pretiosa novella*. In *Bibliotheca chemica curiosa*, 2:1–80.

Philalethes, Eirenaeus [George Starkey]. *Introitus apertus ad occlusum regis palatium*. In *Museum hermeticum*, pp. 647–99.

————. *Ripley Reviv'd*. London, 1678.

————. *Secrets Reveal'd; or, An Open Entrance to the Shut-Palace of the King*. London, 1669.

Pike, Albert. *Morals and Dogma of the Ancient and Accepted Scottish Rite*. London, 1871.

Plessner, Martin. "Hermes Trismegistus and Arab Science." *Studia Islamica* 2 (1954): 45–59.

————. "Neue Materialien zur Geschichte der Tabula Smaragdina." *Der Islam* 16 (1928): 77–113.

————. "The Place of the *Turba Philosophorum* in the Development of Alchemy." *Isis* 45 (1954): 331–38.

————. *Vorsokratische Philosophie und griechische Alchemie*. Wiesbaden: Steiner, 1975.

Pluche, Noël Antoine. *Histoire du ciel*. 2 vols. Paris, 1757.

Poisson, Albert. *Théories et symboles des alchimistes*. Paris, 1891.

Porto, Paulo Alves. "'Summus atque felicissimus salium': The Medical Relevance of the Liquor Alkahest." *Bulletin of the History of Medicine* 76 (2002): 1–29.

Post, Gaines. "Master's Salaries and Student-Fees in Mediaeval Universities." *Speculum* 7 (1932): 181–98.

Post, Gaines, Kimon Giocarinis, and Richard Kay. "The Medieval Heritage of a Humanistic Ideal: 'Scientia donum dei est, unde vendi non potest.'" *Traditio* 11 (1955): 195–234.

Powers, John C. "'Ars sine Arte': Nicholas Lemery and the End of Alchemy in Eighteenth-Century France." *Ambix* 45 (1998): 163–89.

————. *Inventing Chemistry: Herman Boerhaave and the Reform of the Chemical Arts*. Chicago: University of Chicago Press, 2012.

Prescher, Hans. *Georgius Agricola: Persönlichkeit und Wirken für den Bergbau und das Hüttenwesen des 16. Jahrhunderts*. Weinheim: VCH, 1985.

Price, James. *An Account of some Experiments on Mercury, Silver and Gold, made in Guildford in May, 1782*. Oxford, 1782.

Priesner, Claus. "Johann Thoelde und die Schriften des Basilius Valentinus." In Meinel, *Die Alchemie in der europäischen Kultur- und Wissenschaftgeschichte*, pp. 107–18.

Principe, Lawrence M. "Alchemy Restored." *Isis* 102 (2011): 305–12.

———. "Apparatus and Reproducibility in Alchemy," In *Instruments and Experimentation in the History of Chemistry*, edited by Frederic L. Holmes and Trevor Levere, pp. 55–74. Cambridge, MA: MIT Press, 2000.

———. *The Aspiring Adept: Robert Boyle and His Alchemical Quest*. Princeton, NJ: Princeton University Press, 1998.

———. "Chemical Translation and the Role of Impurities in Alchemy: Examples from Basil Valentine's *Triumph-Wagen*." *Ambix* 34 (1987): 21–30.

———, ed. *Chymists and Chymistry*. Sagamore Beach, MA: Chemical Heritage Foundation and Science History Publications, 2007.

———. "D. G. Morhof's Analysis and Defence of Transmutational Alchemy." In *Mapping the World of Learning: The Polyhistor of Daniel Georg Morhof*, edited by Françoise Wacquet, pp. 138–53. Wolfenbüttler Forschungen 91. Harrassowitz: Wiesbaden, 2000.

———. "Diversity in Alchemy: The Case of Gaston 'Claveus' DuClo, a Scholastic Mercurialist Chrysopoeian." In *Reading the Book of Nature: The Other Side of the Scientific Revolution*, edited by Allen G. Debus and Michael Walton, pp. 181–200. Kirksville, MO: Sixteenth Century Press, 1998.

———. "Georges Pierre des Clozets, Robert Boyle, the Alchemical Patriarch of Antioch, and the Reunion of Christendom: Further New Sources." *Early Science and Medicine* 9 (2004): 307–20.

———. "Reflections on Newton's Alchemy in Light of the New Historiography of Alchemy." In *Newton and Newtonianism: New Studies*, edited by James E. Force and Sarah Hutton, pp. 205–19. Dordrecht: Kluwer, 2004.

———. "Revealing Analogies: The Descriptive and Deceptive Roles of Sexuality and Gender in Latin Alchemy." In *Hidden Intercourse: Eros and Sexuality in the History of Western Esotericism*, edited by Wouter J. Hanegraaff and Jeffrey J. Kripal, pp. 208–29. Leiden: Brill, 2008.

———. "A Revolution Nobody Noticed? Changes in Early Eighteenth Century Chymistry." In *New Narratives in Eighteenth-Century Chemistry*, edited by Lawrence M. Principe, pp. 1–22. Dordrecht: Springer, 2007.

———. *The Scientific Revolution: A Very Short Introduction*. Oxford: Oxford University Press, 2011.

———. "Transmuting Chymistry into Chemistry: Eighteenth-Century Chrysopoeia and Its Repudiation." In *Neighbours and Territories: The Evolving Identity of Chemistry*, edited by José Ramón Bertomeu-Sánchez, Duncan Thorburn Burns, and Brigitte Van Tiggelen, pp. 21–34. Louvain-la-Neuve, Belgium: Mémosciences, 2008.

———. "Van Helmont." In *Dictionary of Medical Biography*, edited by W. F. Bynum and Helen Bynum, 3:626–28. Westport, CT: Greenwood Press, 2006.

———. *Wilhelm Homberg and the Transmutations of Chymistry*. Forthcoming.

Principe, Lawrence M., and Lloyd Dewitt. *Transmutations: Alchemy in Art*. Philadelphia: Chemical Heritage Foundation, 2002.

Principe, Lawrence M., and William R. Newman. "Some Problems in the Historiography of Alchemy." In *Secrets of Nature: Astrology and Alchemy in Early Modern Europe*, edited by William Newman and Anthony Grafton, pp. 385–434. Cambridge, MA: MIT Press, 2001.

Prinke, Rafal T. "Beyond Patronage: Michael Sendivogius and the Meanings of Success in Alchemy." In López-Perez, *Chymia*, pp. 175–231.

Pumphrey, Stephen. "The Spagyric Art; or, The Impossible Work of Separating Pure from Impure Paracelsianism: A Historiographical Analysis." In Grell, *Paracelsus*, pp. 21–51.

Putscher, Marielene. "Das *Buch der heiligen Dreifaltigkeit* und seine Bilder in Handschriften des 15. Jahrhunderts." In Meinel, *Die Alchemie in der europäischen Kultur- und Wissenschaftsgeschichte*, pp. 151–78.

Rampling, Jennifer. "The Alchemy of George Ripley, 1470–1700." PhD diss., Clare College, University of Cambridge, 2000.

———. "The Catalogue of the Ripley Corpus: Alchemical Writings Attributed to George Ripley." *Ambix* 57 (2010): 125–201.

———. "Establishing the Canon: George Ripley and His Alchemical Sources." *Ambix* 55 (2008): 189–208.

Ranking, G. S. A. "The Life and Works of Rhazes (Abu Bakr Muhammad bin Zakariya ar-Razi)." *XVII International Congress of Medicine, London 1913, Proceedings*, sec. 23, pp. 237–68.

Rashed, Roshdi, and Régis Morelon, eds. *Histoire des sciences arabes*. Vol. 3, *Technologie, alchimie et sciences de la vie*. Paris: Seuil, 1997.

Rattansi, Piyo, and Antonio Clericuzio, eds. *Alchemy and Chemistry in the Sixteenth and Seventeenth Centuries*. Dordrecht: Kluwer, 1994.

Ray, Praphulla Chandra. *A History of Hindu Chemistry*. 2 vols. London: Williams and Norgate, 1907–9. Expanded ed., under the title *History of Chemistry in Ancient and Medieval India*, Calcutta: Indian Chemical Society, 1956.

Read, John. *Prelude to Chemistry: An Outline of Alchemy, Its Literature and Relationships*. London: Bell and Sons, 1936.

Rebotier, Jacques. "La musique cachée de l'*Atalanta fugiens*." *Chrysopoeia* 1 (1987): 56–76.

———. "La *Musique de Flamel*." In Kahn and Matton, *Alchimie: Art, histoire, et mythes*, pp. 507–46.

Regardie, Israel. *The Philosopher's Stone: A Modern Comparative Approach to Alchemy from the Psychological and Magical Points of View*. London: Rider, 1938.

Reidy, J. "Thomas Norton and the *Ordinall of Alchimy*." *Ambix* 6 (1957): 59–85.

Rey Bueno, Mar. "La alquimia en la corte de Carlos II (1661–1700)." *Azogue* 3 (2000). Online at http://www.revistaazogue.com.

———. *Los señores del fuego: Destiladores y espagíricos en la corte de los Austrias*. Madrid: Corona Borealis, 2002.

Reyher, Samuel. *Dissertatio de nummis quibusdam ex chymico metallo factis*. Kiel, Germany, 1690.

Ricketts, Mac Linscott. *Mircea Eliade: The Romanian Roots, 1907–1945*. Boulder, CO: East European Monographs, 1988.

Ripley, George. *Compound of Alchymie*. In Ashmole, *Theatrum Chemicum Britannicum*, 107–93.

Roosen-Runge, Heinz. *Farbgebung und Technik frümittelalterlicher Buchmalerei: Studien zu den Traktaten "Mappae Clavicula" und "Heraclius."* 2 vols. Munich: Deutscher Kunstverlag, 1967.

Rosarium philosophorum: Ein alchemisches Florilegium des Spätmittelalters. Edited by Joachim Telle. 2 vols. Weinheim: VCH, 1992.

Rose, Thomas Kirke. "The Dissociation of Chloride of Gold." *Journal of the Chemical Society* 67 (1895): 881–904.

Ruff, Andreas. *Die neuen kürzeste und nützlichste Scheide-Kunst oder Chimie theoretisch und practisch erkläret.* Nuremberg, 1788.

Ruska, Julius. "Al-Biruni als Quelle für das Leben und die Schriften al-Rāzī's." *Isis* 5 (1923): 26–50.

———. "Die Alchemie ar-Razi's." *Der Islam* 22 (1935): 281–319.

———. "Die Alchemie des Avicenna." *Isis* 21 (1934): 14–51.

———. *Al-Rāzī's Buch der Geheimnis der Geheimnisse.* Berlin: Springer, 1937. Reprint, Graz: Verlag Edition Geheimes Wissen, 2007.

———. *Arabische Alchemisten I: Chālid ibn-Jazīd ibn- Mu'āwija. Heidelberger Akten von-Portheim-Stiftung* 6 (1924). Reprint, Vaduz, Liechtenstein: Sändig Reprint Verlag, 1977.

———. *Arabische Alchemisten II: Ğa'far alṢādiq, der Sechste Imām. Heidelberger Akten von-Portheim-Stiftung* 10 (1924). Reprint, Vaduz, Liechtenstein: Sändig Reprint Verlag, 1977.

———. *Tabula Smaragdina: Ein Beitrag zur Geschichte der hermetischen Literatur.* Heidelberg: Winter, 1926.

———. *Turba philosophorum: Ein Beitrag zur Geschichte der Alchemie.* Berlin: Springer, 1931.

Ruska, Julius, and E. Wiedemann. "Beiträge zur Geschichte der Naturwissenschaften LXVII: Alchemistische Decknamen." *Sitzungsberichte der Physikalisch-medizinalischen Societät zu Erlangen* 56 (1924): 17–36.

S. A. [Sapere Aude, pseudonym of William Wynn Westcott]. *The Science of Alchymy.* London: Theosophical Publishing Society, 1893.

Saffrey, Henri Dominique. "Historique et description du manuscrit alchimique de Venise *Marcianus graecus* 299." In Kahn and Matton, *Alchimie: Art, histoire, et mythes,* Textes et Travaux de Chrysopoeia 1, pp. 1–10. Paris: SÉHA; Milan: Archè, 1995.

Sala, Angelo. *Processus de auro potabili.* Strasbourg, 1630.

Schott, Heinz, and Ilana Zinguer, eds. *Paracelsus und seine internationale Rezeption in der frühen Neuzeit.* Leiden: Brill, 1998.

Segonds, Alain-Philippe. "Astronomie terrestre/Astronomie céleste chez Tycho Brahe." In *Nouveau ciel, nouvelle terre: La révolution copernicienne dans l'allemagne de la réforme (1530–1630),* edited by Miguel Ángel Granada and Édouard Mehl, pp. 109–42. Paris: Les Belles Lettres, 2009.

———. "Tycho Brahe et l'alchimie." In Margolin and Matton, *Alchimie et philosophie à la Renaissance,* pp. 365–78.

Shackelford, Jole. "Tycho Brahe, Laboratory Design, and the Aim of Science: Reading Plans in Context." *Isis* 84 (1993): 211–30.

Siggel, Alfred. *Decknamen in der arabischen alchemistischen Literatur.* Berlin: Akademie Verlag, 1951.

Silberer, Herbert. *Hidden Symbolism of Alchemy and the Occult Arts.* New York: Dover, 1971.

Sivin, Nathan. *Chinese Alchemy: Preliminary Studies.* Cambridge, MA: Harvard University Press, 1968.

———. "Research on the History of Chinese Alchemy." In *Alchemy Revisited,* edited by Z. R. W. M. von Martels, pp. 3–20. Leiden: Brill, 1990.

Slater, John. "Rereading Cabriada's *Carta:* Alchemy and Rhetoric in Baroque Spain." *Colorado Review of Hispanic Studies* 7 (2009): 67–80.

Smith, Cyril Stanley, and John G. Hawthorne. *Mappae Clavicula: A Little Key to the World of Medieval Techniques.* Transactions of the American Philosophical Society 64. Philadelphia: American Philosophical Society, 1974.

Smith, Pamela H. "Alchemy as a Language of Mediation in the Habsburg Court." *Isis* 85 (1994): 1–25.

———. *The Business of Alchemy: Science and Culture in the Holy Roman Empire.* Princeton, NJ: Princeton University Press, 1994.

Stahl, Georg Ernst. *Fundamenta chymiae dogmaticae.* Leipzig, 1723.

———. *Philosophical Principles of Universal Chemistry.* Translated by Peter Shaw. London, 1730.

Stapleton, H. E., R. F. Azo, and M. Hidayat Husain. "Chemistry in Iraq and Persia in the Tenth Century AD." *Memoirs of the Asiatic Society of Bengal* 8 (1927): 317–418.

Stapleton, H. E., R. F. Azo, Hidayat Husain, and G. L. Lewis. "Two Alchemical Treatises Attributed to Avicenna." *Ambix* 10 (1962): 41–82.

Starkey, George. *The Alchemical Laboratory Notebooks and Correspondence of George Starkey.* Edited by William R. Newman and Lawrence M. Principe. Chicago: University of Chicago Press, 2004.

———. *Liquor Alkahest.* London, 1675.

Steele, Robert B. "The Treatise of Democritus on Things Natural and Mystical." *Chemical News* 61 (1890): 88–125.

Stoltzius von Stoltzenberg, Daniel. *Chymisches Lustgärtlein.* Frankfurt, 1624. Reprint, Darmstadt: Wissenschaftliche Buchgesellschaft, 1964.

Stolzenberg, Daniel. "Unpropitious Tinctures: Alchemy, Astrology, and Gnosis according to Zosimos of Panopolis." *Archives internationales d'histoire des sciences* 49 (1999): 3–31.

Stone of the Philosophers. In *Collectanea chymica*, pp. 55–120.

Strohmaier, Gotthard. "Al-Mansūr und die frühe Rezeption der griechischen Alchemie." *Zeitschrift für Geschichte der Arabisch-Islamischen Wissenschaften* 5 (1989): 167–77.

———. "'Umāra ibn Hamza, Constantine V, and the Invention of the Elixir." *Graeco-Arabica* 4 (1991): 21–24.

Sutherland, C. H. V. "Diocletian's Reform of the Coinage: A Chronological Note." *Journal of Roman Studies* 45 (1955): 116–18.

Tachenius, Otto. *Epistola de famoso liquore alcahest.* Venice, 1652.

———. *Hippocrates chymicus.* London, 1677.

Tanckius, Joachim. *Promptuarium Alchemiae.* 2 vols. Leipzig, 1610 and 1614. Reprint, Graz: Akademische Druck, 1976.

Taylor, Frank Sherwood. "Alchemical Works of Stephanus of Alexandria, Part I." *Ambix* 1 (1937): 116–39.

———. "Alchemical Works of Stephanus of Alexandria, Part II." *Ambix* 2 (1938): 39–49.

———. *The Alchemists: Founders of Modern Chemistry.* New York: Schuman, 1949.

Telle, Joachim, ed. *Analecta Paracelsica: Studien zum Nachleben Theophrast von Hohenheims im deutschen Kulturgebiet der frühen Neuzeit.* Stuttgart: Franz Steiner Verlag, 1994.

———. "Chymische Pflanzen in der deutschen Literatur." *Medizinhistorisches Journal* 8 (1973): 1–34.

———. "Paracelsistische Sinnbildkunst: Bemerkungen zu einer Pseudo-*Tabula smaragdina* des 16. Jahrhunderts." In *Bausteine zur Medizingeschichte*, pp. 129–39. Wiesbaden: Franz Steiner Verlag, 1984. French translation: "L'art symbolique paracelsien: Remarques concernant une pseudo-*Tabula smaragdina* du XVIᵉ siècle." In *Présence de Hèrmes Trismégeste*, edited by Antoine Faivre, pp. 184–208. Paris: Albin Michel, 1988.

———. "Remarques sur le *Rosarium philosophorum* (1550)." *Chrysopoeia* 5 (1992–96): 265–320.

Theatrum chemicum. 6 vols. Strasbourg, 1659–63. Reprint, Torino: Bottega d'Erasmo, 1981.

Theophilus. *On Divers Arts*. Translated by John G. Hawthorne and Cyril Stanley Smith. New York: Dover, 1979.

Tiffereau, Cyprien Théodore. *L'art de faire l'or*. Paris, 1892.

———. *Les métaux sont des corps composés*. Vaugirard, 1855. Reprinted as *L'or et la transmutation des métaux* (Paris, 1889).

Travaglia, Pinella. "I *Meteorologica* nella tradizione eremetica araba: Il *Kitāb sirr al balīqa*." In Viano, *Aristoteles chemicus*, pp. 99–112.

———. *Magic, Causality and Intentionality: The Doctrine of Rays in al-Kindī*. Micrologus Library 3. Florence: Sismel, 1999.

Ullmann, Manfred. "Hālid ibn-Yazīd und die Alchemie: Eine Legende." *Der Islam* 55 (1978): 181–218.

———. *Die Natur- und Geheimwissenschaften im Islam*. Leiden: Brill, 1972.

Valentine, Basil. *Chymische Schrifften*. 2 vols. Hamburg, 1677. Reprint, Hildesheim: Gerstenberg Verlag, 1976.

———. *Ein kurtz summarischer Tractat . . . von dem grossen Stein der Uhralten*. Eisleben, 1599.

Van Bladel, Kevin T. *The Arabic Hermes: From Pagan Sage to Prophet of Science*. Oxford: Oxford University Press, 2009.

Van Helmont, Joan Baptista. *Opuscula medica inaudita*. Amsterdam, 1648. Reprint, Brussels: Culture et Civilization, 1966.

———. *Ortus medicinae*. Amsterdam, 1648. Reprint, Brussels: Culture et Civilization, 1966.

Ventura, Lorenzo. *De ratione conficiendi lapidis philosophici*. In *Theatrum chemicum*, 2:215–312.

Viano, Cristina, ed. *L'Alchimie et ses racines philosophiques: La tradition grecque et la tradition arabe*. Paris: Vrin, 2005.

———. "Les alchimistes gréco-alexandrins et le *Timée* de Platon." In Viano, *L'Alchimie et ses racines philosophiques*, pp. 91–108.

———. "Aristote et l'alchimie grecque." *Revue d'histoire des sciences* 49 (1996): 189–213.

———, ed. *Aristoteles chemicus: Il IV libro dei* Meteorologica *nella tradizione antica e medievale*. Sankt Augustin, Germany: Academia Verlag, 2002.

———. "Gli alchimisti greci e l'acqua divina." *Rendiconti della Accademia Nazionale delle Scienze. Parte II: Memorie di scienze fisiche e naturali* 21 (1997): 61–70.

———. *La matière des choses: Le livre IV des Météorologiques d'Aristote et son interprétation par Olympiodore*. Paris: Vrin, 2006.

———. "Olympiodore l'alchimiste et les Présocratiques." In Kahn and Matton, *Alchimie: Art, histoire, et mythes*, pp. 95–150.

Vinciguerra, Antony. "The *Ars alchemie:* The First Latin Text on Practical Alchemy." *Ambix* 56 (2009): 57–67.

Waddell, Mark A. "Theatres of the Unseen: The Society of Jesus and the Problem of the Invisible in the Seventeenth Century." PhD diss., Johns Hopkins University, 2006.

Waite, Arthur Edward. *Azoth; or, The Star in the East, Embracing the First Matter of the Magnum Opus, the Evolution of the Aphrodite-Urania, the Supernatural Generation of the Son of the Sun, and the Alchemical Transfiguration of Humanity.* London, 1893. Reprint, Secaucus, NJ: University Books, 1973.

———. *Lives of the Alchemystical Philosophers.* London, 1888. Reprinted under the title *Alchemists through the Ages,* New York: Rudolf Steiner Publications, 1970.

———. *The Secret Tradition of Alchemy.* New York: Alfred Knopf, 1926.

Wamberg, Jacob, ed. *Alchemy and Art.* Copenhagen: Museum Tusculanum Press, 2006.

Warlick, M. E. *Max Ernst and Alchemy: A Magician in Search of a Myth.* Austin: University of Texas Press, 2001.

Wedel, Georg Wolfgang. "Programma vom Basilio Valentino." In *Deutsches Theatrum Chemicum,* 1:669–80.

Weisser, Ursula. *Das "Buch über das Geheimnis der Schöpfung" von Pseudo-Apollonios von Tyana.* Berlin: Walter de Gruyter, 1980. Reprint de Gruyter, 2010.

Westcott, William Wynn. *See* S. A. [Sapiere, Aude].

Westfall, Richard S. "Alchemy in Newton's Library." *Ambix* 31 (1994): 97–101.

Weyer, Jost. *Graf Wolfgang von Hohenlohe und die Alchemie: Alchemistische Studien in Schloss Weikersheim 1587–1610.* Sigmaringen, Germany: Thorbecke, 1992.

Wiedemann, Eilhard. "Zur Alchemie bei der Arabern." *Journal für praktische Chemie* 184 (1907): 115–23.

Wiegleb, Johann Christian. *Historisch-kritische Untersuchung der Alchimie.* Weimar, 1777. Reprint, Leipzig: Zentral-Antiquariat der DDR, 1965.

Wieland, Christoph Martin. "Der Goldmacher zu London." *Teutsche Merkur,* February 1783, pp. 163–91.

Wilsdorf, H. M. *Georg Agricola und seine Zeit.* Berlin: Deutsche Verlag der Wissenschaften, 1956.

Winter, Alison. *Mesmerized: Powers of Mind in Victorian Britain.* Chicago: University of Chicago Press, 1998.

Wujastyk, Dominik. "An Alchemical Ghost: The Rasaratnakara by Nagarjuna." *Ambix* 31 (1984): 70–84.

Zanier, Giancarlo. "Procedimenti farmacologici e pratiche chemioterapeutiche nel *De consideratione quintae essentiae.*" In *Alchimia e medicina nel Medioevo,* edited by Chiara Crisciani and Agostino Paravicini Bagliani, pp. 161–76. Micrologus Library 9. Florence: Sismel, 2003.

Ziegler, Joseph. *Medicine and Religion c. 1300: The Case of Arnau de Vilanova.* Oxford: Clarendon Press, 1998.

Zosimos of Panopolis. *On the Letter Omega.* Edited and translated by Howard M. Jackson. Missoula, MT: Scholars Press, 1978.